Long-Lived States in Collisions

Author

Slobodan D. Bosanac
The Rugjer Bošković Institute
Zagreb, Croatia
Yugoslavia

CRC Press
Taylor & Francis Group
Boca Raton London New York

CRC Press is an imprint of the
Taylor & Francis Group, an **informa** business

First published 1988 by CRC Press
Taylor & Francis Group
6000 Broken Sound Parkway NW, Suite 300
Boca Raton, FL 33487-2742

Reissued 2018 by CRC Press

Library of Congress Cataloging-in-Publication Data

Bosanac, Slobodan D.
Long-lived states in collisions.
Bibliography: p.
Includes index.
1. Collisions (Nuclear physics) I. Title.
QC794.6.C6B67 1988 539.7'54 87-17845
ISBN 0-8493-6871-5

A Library of Congress record exists under LC control number: 87017845

Publisher's Note
The publisher has gone to great lengths to ensure the quality of this reprint but points out that some imperfections in the original copies may be apparent.

ISBN 13: 978-1-315-89506-2 (hbk)
ISBN 13: 978-1-351-07416-2 (ebk)

Visit the Taylor & Francis Web site at http://www.taylorandfrancis.com and the
CRC Press Web site at http://www.crcpress.com

To my mother

PREFACE

The idea of writing a book about the long-lived states in collisions was brewing for some time, but starting it came quite suddenly. It was a coincidence of two events. One was that new ideas developed in understanding of such states, and the other was a discussion with a friend, Prof. Nenad Trinajstić, about some aspects of these states, after which he suggested that I write a book about all these. It was a challenge that I accepted, not realizing what was awaiting me.

First, I had to convince myself that such a book is really needed. The importance of long-lived states in collisions cannot be disputed. In fact, some of the most interesting processes have such states as their intermediary. For example, large numbers of long-lived states appear in collision of stable elementary particles, e.g., protons, electrons, positrons, etc., and for some specific reasons they are also called elementary particles. Some of them live for such a long time that they even leave a clearly visible trace in, say, a bubble chamber. An alpha-decaying nucleus is also an example of a long-lived state, but it is treated as a decay process rather than as an intermediary state in collision. Less obvious but a very important example of the long-lived states is connected with the time irreversible processes in systems with large number degrees of freedom, such as a glass of water. Usually such systems are very successfully treated by the statistical theories, but these theories do not answer the basic question: What causes the time irreversibility of the processes in such systems, when it is known that the equations of motion for atoms and molecules are time-reversible? Again, if one looks in details, what seems to be the time irreversible process is in fact manifestation of the properties of the long-lived states which are formed in collisions of large clusters of particles.

These scattered examples only show that one finds long-lived states on different microscopic levels. It is not clear, however, that in all these cases the long-lived states may have some common properties. One common feature in those examples is that the systems are quantum, i.e., their properties must be described by the quantum theory. The classical theory is seldom good and only for the qualitative understanding of these systems. However, in the quantum theory of scattering there is a phenomenon called resonance which, it is argued, represents a long-lived state and therefore, in principle, all the above-mentioned long-lived states can be described by these resonances. There are numerous books which describe resonances, and therefore adding this one to an already big pile would just be repeating what is already known about them, or at best it would be applying this knowledge to atom and molecule collisions. Whether this is sufficiently interesting to justify writing a book is questionable. However, there is a small problem with the long-lived states which haunted me for long time, but I thought it could be solved in terms of resonances, although it was not obvious to me how. Imagine that we collide a particle with a large cluster of particles (say an atom with a large molecule). If the forces between these two species are attractive from their infinite until a relatively small separation (which is the case with atoms and molecules) then on impact with the cluster the incoming particle may be able to transfer most of its kinetic energy into vibrations of the cluster, and stay bound to it. This process can be very nicely observed in the classical simulation of collision, and indeed it was described on many occasions for the case of atom-molecule systems. It was further observed that the original particle stays bound to the cluster for a long time, until part of the excess energy is not released by the dissociation process, i.e., by the fragmentation of the cluster. It was also observed that formation of the intermediary long-lived state is not very sensitive on the variation of the incident energy of the particle. In other words, a long-lived state is formed even for a relatively large variation of the collision energy. Yet it is known that resonances have a unique property. The longer they live the narrower the interval of the collision energy within which they are observed. Therefore it appears that the long-lived states, which we described, have properties that are completely opposite to those of resonances. Of course, it is not legitimate to compare these two kinds of long-lived states because the former is a result of the classical simulation while the latter is the quantum phenomenon. However, one can always make a quantum system behave more and more classically, by taking smaller and smaller value of the Planck's constant h. In such a case it is known that the number of resonances, per unit interval of energy, increases and therefore in the end one obtains the classical effect of the long-lived states as the interference effect of many overlapping resonances. Such an explanation cannot be excluded, but it did not look right to me. This was further substantiated by the fact that in the same limit the width of resonances also descreases. Since one cannot make beautiful theories by having only the feeling that something is not right, the idea that there may be other types of long-lived states, besides resonances, was waiting until the proof was available that this is indeed the case. Further research showed that resonances are not long-lived states in a true sense, which, at least to me, was quite intriguing. At this point I thought that the results were interesting enough so that writing a book on the long-lived states would have some justification.

The plan for the book was that it would be useful to scientists other than atom and molecular physicists, and so I tried to make it as general as possible, however, with the emphasis on the atom and molecule collisions. There are some specific features of these collisions which must be taken into account, but they are not of such character that the theory is less general. On the contrary, by considering these features one learns a great deal about the collision theory for the systems with hard core potentials and about the quantum theory in the limit of ''very small h''. In addition, throughout the book we will use the method of the complex angular momentum analysis of the scattering amplitude. In its original form, when this method was introduced in the potential scattering theory, it was applied to systems for which the implicit assumption was that they do not have hard core. Therefore, original theory cannot be used for description of atom and molecule collisions, but with suitable modification it becomes a powerful tool.

The book contains essentially two parts. A review of the classical, quantum, and semi-classical theories of collisions are given in the first part, while their applications to the atom and molecule collisions are given in the second part. The idea in the first part was to give an overview of essential parts in the general theory of collisions which are relevant to the study of the long-lived states. Originally this part was planned to be brief, but in the course of writing it became obvious that very little is known about the topic. When it was finished I realized that this ''would-be introduction'' had taken almost half of the book. It contains what I believe to be the major points in the theory of collisions, with extensive discussions of the drawbacks in the classical and quantum theory. As an illustration of these drawbacks I have discussed simple examples. For some, these examples may appear too trivial to be mentioned, but I hope that they serve the purpose. On the other hand, if they turn out to be failures, I hope that they will make the interested reader think about the better ones. It would be disastrous if these examples led the reader to the wrong conclusions.

In this part of the book I have also included a chapter on decay. These processes are not the subject of the collision theory, but very often it is assumed that decays are ''half'' of an appropriate scattering problem. Intuitively this is an attractive model, especially if a decaying system lives for a long time before it splits into fragments. The model is founded on the belief that the long-lived states ''forget'' their origin and therefore, no matter how they are formed, they will always fragment in one way. However, such a belief is not realistic, as discussed in this chapter. Furthermore, it is shown that there are very few truly decaying processes. Many of them appear to be decays, however, they have their history, even when long-lived states are involved.

Analysis of the long-lived states in atom and molecule collisions is given in the second part of the book. Since there is a large amount of information about simple systems, and as one goes to more complex systems the knowledge about them becomes poorer, the chapters likewise shrink towards the end of the book. Complex systems are difficult to analyze because the relevant equations of motion are difficult to solve. Also there are some fundamental problems, which are not related to our inability of solving the equations of motion, but are connected with the properties of long-lived states in complex systems. The roots of these problems are mentioned in the first part of the book, and they are traced to the immensely complicated analytic structure of the scattering amplitude. Therefore as one goes from simple systems to more complex ones, the methods of analysis change from the quantum to entirely classical theory of scattering. One reason is that for complex systems it is much easier to solve the initial value problem (which is the classical theory) than the boundary value problem (which is the quantum theory). The other reason is that in the quantum theory of scattering one makes expansion of the wave function in the asymptotic states, which is poorly convergent. In the classical theory this problem does not exist, and therefore it is much easier to apply this theory than the quantum. However, the meaning of the results obtained from the classical theory of collisions is questionable, as is partly discussed in the chapter on the semiclassical theory of scattering.

Overall this book is the theoretician's view of the long-lived states in collisions. I did not discuss the experimenal side of this problem or, in particular, the methods how to measure the time delay. This does not mean that I am totally ignorant of the experimental side of scientific research, and for this I am greatly thankful to my friends, Prof. U. Buck and Prof. J. P. Toennies, both from the MPI für Strömungsforschung in Göttingen. They have taught me the necessary rudiments about the experiments in atomic and molecular collisions, which every theoretician should know. I am also thankful to them for many discussions, and long collaboration in this field. As a theoretician I should be thankful to many colleagues from different fields of science, from whom I have learned that the basis of the most important laws in nature is simplicity. From them I have also learned that the beauty of nature lies in a variety of details, which are nothing else but reflections of a few basic laws. I am thankful to Prof. J. N. Murrell from the University of Sussex, who taught me how to be a scientist and to my father Prof. T. Bosanac, with whom I had endless discussions about different problems in science, although very often it was he who did the talking. I would be doing a great injustice if I only thanked those who directly or indirectly contributed to the scientific content of the book. Without the patience and understanding of my wife, Ana, this book would never have seen the daylight. I am grateful to her for this and also for many lonely months when she was waiting for me to return from visits to my colleagues. Our children, Jana and Iva, were too young to understand why their father was very often too busy with science to be able to play with them. I only hope that one day they will forgive me for this.

In the end, I would like to thank my friend Prof. H. Kroto from the University of Sussex, for using free time in his immensely busy schedule to critically review the manuscript and Dr. Z. Bačić from the Institute of Physics in Zagreb, for making valuable comments about the book. Also, thanks go to my secretary, Mrs. B. Špoljar, who painstakingly typed my handwriting which sometimes was even difficult for me to read.

THE AUTHOR

Prof. Slobodan Danko Bosanac, Ph.D., is Science Advisor in the Department of Physical Chemistry at the Rugjer Bošković Institute in Zagreb, Croatia, Yugoslavia.

Prof. Bosanac received his higher education at the University of Zagreb where he graduated from the Department of Theoretical Physics in the field of high energy physics. After graduation, he obtained the SRC grant from the British government to do research at the University of Sussex under the supervision of Prof. J. N. Murrell. During the period between 1969 and 1972 he worked on the problems of bound states of three atom systems using the hyperspherical coordinates. The theory was applied to a particular problem of He trimer. He also worked on the reactive collision problems in connection with vibrational predissociation of carbon dioxide molecules.

In 1972 he obtained a Ph.D. from the University of Sussex and obtained a 2-year SRC postdoctoral grant to work with Dr. G. G. Balint-Kurti at the University of Bristol, England. During this period, he worked on the rotational inelastic problems of simple systems. He returned to Zagreb in 1974 where he continued research at the Rugjer Bošković Institute until 1975 when he was in the Military Service for 1 year.

From 1974 to 1976 he became interested in the theoretical problems of long-lived states in collisions. For this, he generalized the method of complex angular momenta to molecular collisions and developed perturbation theory for the poles of S-matrix. In 1977 he obtained the Lady Davis Fellowship to do research at the Hebrew University of Jerusalem with Prof. R. B. Gerber. During this year, he formulated the semiclassical perturbation theory of collisions, which was necessary for proper treatment of molecular collision problems. Also, the technique of complex angular momentum poles was applied to the inversion problem of scattering.

After his return to Zagreb, Prof. Bosanac investigated various inelastic collision processes in order to understand the mechanism of molecular energy transfer and rules of formation of long-lived states. Models of rotational and vibrational energy transfer were formulated which explained its basic features. He initiated several international projects with the groups at the Max-Planck-Institute for Fluid Dynamics in Göttingen, the University of Sussex, and the University of Florida in Gainesville. He made numerous extended visits to Universities in Europe and the U.S. where he was involved in various projects in the field of molecular energy transfer.

Prof. Bosanac has published over 50 scientific papers related to molecular energy transfer and general topics in mathematical physics.

TABLE OF CONTENTS

Chapter 1

CLASSICAL THEORY OF COLLISIONS

I. GENERAL ASPECTS

Newton's equations of motion form the basis of the classical theory of collision. They are second order nonlinear differential equations in time, and their solution represents the time evolution of the coordinates of all particles. Classical theory is an initial value problem in one variable (time), which means that the evolution of systems of particles is uniquely determined if at some instant of time all their coordinates and velocities are known. The set of all coordinates can be written as a column vector U and will be called a trajectory of the system of particles. The time derivative of such a trajectory is $\mathring{U}$, and thus, Newton's equations are the equations for this trajectory.

Newton's equations can be formulated in various ways. One of them is on the basis of the principle of least action, the Hamilton principle, which says that the integral of a function S, called the Lagrange function, between two points in time has its extreme value only along the classical trajectory. Explicitly, if we write the deviation from the classical trajectory U as the column vector $\epsilon(t)$, then the integral

$$I(\epsilon) = \int_{t_1}^{t_2} S(U + \epsilon, \mathring{U} + \mathring{\epsilon})\, dt \qquad (1)$$

is sensitive to second order in $\epsilon(t)$, i.e.,

$$I(\epsilon) = \int S(U, \mathring{U})\, dt + O(\epsilon^2) \qquad (2)$$

If ϵ is zero at the end points, i.e., $\epsilon(t_1) = \epsilon(t_2) = 0$, it follows, from this principle, that the Lagrange function S satisfies the set of equations

$$\frac{\partial S}{\partial u_i} = \frac{d}{dt}\left(\frac{\partial S}{\partial \mathring{u}_i}\right) \qquad (3)$$

where u_i is the ith element of U. The Lagrange function has the form

$$S = T(U) - V(U) \qquad (4)$$

where T is the kinetic energy and V(U) is the potential. In Cartesian coordinates the kinetic energy for a system of n + 1 particles has the form

$$T = \frac{1}{2}\sum_{i=1}^{n+1} m_i(\mathring{x}_i^2 + \mathring{y}_i^2 + \mathring{z}_i^2) \qquad (5)$$

where m_i is the mass of the ith particle and x_i, y_i, and z_i are the respective three-dimensional (3D) coordinates. The potential is independent of velocities $\mathring{U}$ and time, a necessary requirement for conservative systems, i.e., those which conserve energy. It is also a function of the relative separations of atoms and molecules.

The set of Equations 3 describes the motion of particles relative to an arbitrary coordinate system, the laboratory coordinate system (L). It is convenient, however, to work in the center of mass coordinate system (C), in which the total momentum of the particles is zero. The transformation between these systems is not unique and depends very much on the circumstances. Here we will briefly describe the transformation which includes relative coordinates in the C system. For simplicity, only Cartesian coordinates x_i will be transformed, because the other coordinates, y_i and z_i, are transformed in a similar way.

The relative coordinates are defined as

$$q_i = x_{i+1} - x_i; \quad i = 1, 2, \ldots, n \tag{6}$$

while the coordinate R of the center of mass is

$$R_x = \frac{m_1x_1 + m_2x_2 + \cdots + m_{n+1}x_{n+1}}{m_1 + m_2 + \cdots + m_{n+1}} = \mu_1x_1 + \mu_2x_2 + \cdots + \mu_{n+1}x_{n+1} \tag{7}$$

The coordinates q_i and R_x can be represented by a column vector q, which allows Equations 6 and 7 to be written as $q = Ax$. The inverse transformation is $x = A^{-1}q$ or, more explicitly,

$$\begin{aligned} x_k = {} & R_x + \mu_1q_1 + (\mu_1 + \mu_2)\,q_2 + \cdots + (\mu_1 + \mu_2 + \cdots + \mu_{k-1})\,q_{k-1} \\ & - (\mu_{k+1} + \mu_{k+2} + \cdots + \mu_{n+1})\,q_k - \cdots - \mu_{n+1}q_n; \\ & k = 1, 2, \ldots, n + 1 \end{aligned} \tag{8}$$

The kinetic energy for the x coordinate is now

$$T_x = \frac{1}{2}\mathring{\tilde{x}}\,m\,\mathring{x} = \frac{1}{2}\mathring{\tilde{q}}\,\tilde{A}^{-1}\,m\,A^{-1}\,\mathring{q} \tag{9}$$

where m is the diagonal matrix of the masses and $\tilde{x}$ is transpose of x. In the new coordinates, the kinetic energy T_x is a bilinear form with zero coefficients for the elements $\mathring{q}_i\mathring{R}_x$; $i = 1, 2, \ldots, n$. Therefore, in the Coordinates 6 and 7 the motion of the center of mass R_x is uncoupled from the other coordinates q_i, and since the potential is not a function of R_x, it follows from Equation 3 that the equation of motion for the center of mass is

$$\mathring{\mathring{R}}_x = 0 \tag{10}$$

i.e., the motion is uniform.

The bilinear form (Equation 9) is explicitly

$$\begin{aligned} T_x = {} & \frac{1}{2}M\sum_{i=1}^{n}(\mu_1 + \mu_2 + \cdots + \mu_i)(1 - \mu_1 - \mu_2 - \cdots - \mu_i)\,\mathring{q}_i^2 \\ & + M\sum_{i<j}^{n}(\mu_1 + \mu_2 \cdots + \mu_i)(1 - \mu_1 - \mu_2 - \cdots - \mu_j)\,\mathring{q}_i\mathring{q}_j \\ & + \frac{1}{2}M\,\mathring{R}_x^2 \end{aligned} \tag{11}$$

where $M = m_1 + m_2 + \ldots + m_{n+1}$. The equations of motion for u_i which are obtained from Equation 3 are not suitable for computation. We must transform q in such a way that the bilinear form of the kinetic energy becomes diagonal. This transformation is achieved by a unitary matrix W which has the property

$$\tilde{W}\,\tilde{A}^{-1}\,m\,A^{-1}\,W = \omega \tag{12}$$

where ω is a diagonal matrix. Therefore, if we define new coordinates e by

$$q = W\,e \tag{13}$$

then

$$T_x = \frac{1}{2}\,\tilde{e}\,\omega\,e \tag{14}$$

is diagonal bilinear form.

Similar transformations apply to T_y and T_z. If we designate by f and g the respective transformed coordinates for y and z, then the total kinetic energy is

$$T = \frac{1}{2}\,(\tilde{e}\,\omega\,e + \tilde{f}\,\omega\,f + \tilde{g}\,\omega\,g) \tag{15}$$

and the equations of motion for (e, f, g) are now deduced from Equation 3. They are analogous to the equations of motion for (x_i, y_i, z_i), except that the mass m_i is replaced by ω_i.

Whether one uses the L equations of motion in the coordinates (x_i, y_i, z_i) or the C equations in the coordinates (e, f, g) depends very much on the circumstances. From now on we will work with the C equations, which are explicitly given as

$$\omega_i\,\ddot{e}_i = -\frac{\partial V}{\partial e_i};\quad \omega_i\,\ddot{i}_i = -\frac{\partial V}{\partial f_i};\quad \omega_i\ddot{g}_i = -\frac{\partial V}{\partial V g_i} \tag{16}$$

but often a different notation for the coordinates may be necessary.

From the set of Equations 16 we want to obtain as much information as possible on the atom and molecule collisions. There are some specific features of these systems, however, which must be taken into account, but we will describe them in the course of our discussions. At the moment, we will describe a typical atom and molecule collision experiment.

Initially, two molecules A and B (the single atom is also treated as one atom molecule) are well separated so that their interaction is zero. In such a case molecules A and B can be treated as particles with total masses m_A and m_B, respectively, and the separation between their centers of mass is the distance r. Their initial relative velocity is v_i, and by convention in the C system the molecule A travels parallel to the Z-axis from negative towards positive values. Likewise, molecule B travels parallel to the Z-axis in the opposite direction. The lines of these two velocity vectors do not meet, in general, but are shifted by an amount b called, by convention, the impact parameter. The initial separation of the molecules A and B is not known accurately, even in principle. This is because of the uncertainty principle which says that the error in determining the initial momentum μv_i (where $\mu = m_A m_B/[m_A + m_B]$ the reduced mass of the two molecules) is always greater or at least equal to the ratio of the h and the error in determining the initial separation. This inequality is written as

$$\mu\,\Delta\,v_i \geqslant h/\Delta r_i \tag{17}$$

In our case we have implicitly assumed that $\Delta v_i = 0$, and therefore, r_i is, according to Equation 17, totally undetermined. Usually the error tolerance Δv_i and Δr_i is finite, but both can be made small if μ is large, which means that an exact trajectory may only be obtained for infinite mass, i.e., for $\mu \rightarrow \infty$, or equivalently, in the limit $h \rightarrow 0$. We will come back to this principle in the study of the long-lived states.

For an exact knowledge of a trajectory we also require the initial coordinates and velocities of atoms in the molecules A and B. Our previous discussion also applies to these initial conditions and therefore only one half of them can be determined with certainty. However, the practical difficulties in measuring them are so great that we can assume that they are not known. Instead of coordinates and velocities there is alternative information on the initial state of the system, which is more accessible to measurement. The internal energies of molecules A and B, i.e., the rotational, vibrational, and electronic, can be utilized. Also, in principle, one can assume that orientations of the molecules A and B are known with respect to the chosen coordinate system. All this information, however, is not sufficient for a unique specification of the initial conditions necessary for the integration of the Equations 16. In principle we know only half of the required number of the initial conditions, the other half being undetermined due to the uncertainty principle. It will be assumed that in collision only two molecules are produced, e.g., C and D. As in the case of A and B, we can define the relative separation r_f and velocity v_f of the molecules C and D. All final internal coordinates and velocities of these molecules cannot be determined, even in principle. Instead we can only determine the parameters which specified the intial state, i.e., only one half of the total number of parameters necessary for the complete specification of the outgoing trajectory. It is just this aspect of atom and molecule collisions which makes them so specific, from the point of view of the classical theory where everything is completely determined. Working with this limited information set about the system inevitably results in introducing the concept of probability into collision theory.

Let us describe in more detail the collision $A + B \rightarrow C + D$. If the total number of particles (atoms) which is involved in this collision is n, then for a complete specification of the trajectory we have to know $6(n - 1)$ initial conditions in the C system. As was observed earlier, only some of them, at most $3(n - 1)$, can be determined. Let us designate the set of determined initial conditions by Ω_i and those undetermined, by Q_i. Because Q_i are undetermined there is a certain arbitrariness in choosing them, from the range encompassed by their definition, which is imposed by the integrals of motion.

As was explained earlier the sets Ω_i (the elements of this set are called the canonical variables) and Q_i (the elements are called the conjugate canonical variables or the conjugate variables) do not coincide with the coordinates e, f, and g and their time derivatives. However, this does not mean that the members of the set Ω_i are easy to determine. One thing about them is obvious; they split into three subsets. One subset characterizes molecule A, the second characterized molecule B, while the third characterizes the relative motion of the two molecules. The last set is simple; it consists of the C kinetic energies of the molecules A and B along the three axis, which in our coordinate system are $(0, 0, E_r)$. The other two sets are a little more difficult to specify. If the molecules A and B consist of coupled harmonic oscillators, then for the canonical variables we can take energies of the normal modes. If, however, the molecules A and B are not harmonic oscillators, then the problem becomes much more difficult. As yet there is no unique way of characterizing such molecules, in general, using canonical variables which are measurable. The exception is a diatomic molecule for which a canonical variable can be defined, though it should be noted that a diatomic molecule is not representative of a molecule consisting of coupled oscillators. For the moment, however, we will assume that all measurable canonical variables are known, i.e., we know the elements of the set Ω_i, and we will set aside the problem of their determination. Principally, this is a bound state problem, and it is an area of research where slightly different problems arise.

Likewise, determination of conjugate canonical variables is difficult for triatomic and larger molecules, but as before, we will assume that they are known. Both sets Ω_i and Q_i are related to the set of coordinates (e, f, g) and ($\mathring{e}$, $\mathring{f}$, $\mathring{g}$) in general by a nonlinear transformation, which must be unique in both directions, i.e., from (e, f, g) and ($\mathring{e}$, $\mathring{f}$, $\mathring{g}$) to Ω_i and Q_i and vice versa. The strategy for computation is then to fix both Ω_i and Q_i and obtain (e, f, g) and ($\mathring{e}$, $\mathring{f}$, $\mathring{g}$). From the last set we obtain initial conditions for the Equations 16 which are integrated until dissociation products are obtained. In principle, we could also set up equations for the canonical set Ω_i and its conjugate counterpart Q_i, but this may be very cumbersome in practice, especially for reactive collision, i.e., a collision in which parent molecules differ from the product molecules by, for example, exchange of constituent atoms. It is much easier to apply the previously described strategy.

In the final state C + D we can also measure analogous parameters, which we designate by Ω_f. If is obvious that for the fixed values of Ω_i the parameters Ω_f depend on the conjugate variables Q_i. Since in a typical molecular collision experiment both Ω_i and Ω_f are specified, the role of the classical theory is to determine those variables Q_i which connect the two sets. In general, there are several Q_i which connect $\Omega_i \rightarrow \Omega_f$, and they are found by integrating the set of Equations 16.

If the initial parameters Q_i are varied, the final set Ω_f changes. Let us designate by dQ_i a small volume element around Q_i. This volume element has an obvious parametrization $dQ_i = b\ d\ b\ d\phi\ dq_i$, where ϕ is the azimuthal angle of the incident molecule A. It does not contain dz because scattering is independent of the choice of the initial z coordinates of the molecules A and B.

We will now define the cross-section σ which is obtained from the conservation of the total number of particles. If the intensity of the incident beam per dQ_i is unity, then in time dt the total number of particles crossing the area $dS = bdb\ d\phi$ is dt dQ_i. After collision, the intensity of particles per $d\Omega_f$ is then σ, and in time dt the total number of particles in the solid angle $d\Omega$ is dt $d\Omega_f\sigma$. Therefore,

$$b\ d\ b\ d\phi\ dq_i = \sigma\ d\Omega\ d\omega_f \tag{18}$$

or

$$\sigma_{\omega_i \rightarrow \omega_f} = \frac{b}{\left|\frac{d\theta}{db}, \frac{d\omega_f}{dq_i}\right| \sin\theta} \tag{19}$$

where the sets ω_i and ω_f characterize the internal states of the pair of molecules A and B are C + D, respectively, such as their vibrational and rotational energies. If there are several distinct sets q_i which connect $\omega_i \rightarrow \omega_f$ then σ is generalized to

$$\sigma_{\omega_i \rightarrow \omega_f} = \frac{1}{\sin\theta} \Sigma \frac{b}{\left|\frac{d\omega_f}{dq_i}, \frac{d\theta}{db}\right|} \tag{20}$$

where the sum extends over all the sets.

The role of classical theory is, therefore, to find the relationship $\Omega_f(Q_i)$ and thus the ratio in Equation 20, which is related to the Jacobian defined by

$$\frac{dx}{dy} = \det\left|\frac{\partial x_m}{\partial y_n}\right| \tag{21}$$

The knowledge of the ratio $d\omega_f/dq_i$ enables us to calculate the cross-sections, from which most of our knowledge about atomic and molecular collision comes. However, the cross-section is not the only measurable quantity in a collision. Additional information can also be obtained by measuring the so-called time delay, which is designated by τ. This quantity, which is central to our discussion, has meaning only for a single trajectory and is defined as the difference between the time of the trajectory from the initial to the final points and the time $r_i/v_i + r_f/v_f$. This latter corresponds to the time of flight of free molecules. Therefore a zero time delay indicates that there is no interaction between molecules A and B and also between C and D. This is a hypothetical situation because without interaction between molecules A and B the molecules C and D would not be created.

The time delay can be both negative and positive. A negative time delay means that the molecules C and D get to the final point before the molecules A and B travel the distance r_i and molecules C and D the distance r_f, without interaction. This occurs if the molecules acquire a large velocity or the potential between A and B has a strongly repulsive core along the trajectory A + B → C + D. In such a case the effective distance between the initial and final points is shortened. In either case the molecules C and D leave the interaction region sooner than they would have done had there been no interaction between A and B. There is an obvious lower limit to the negative time delay, because a very negative τ value would mean that the molecules C and D leave before the molecules A and B arrive in the interaction region, which obviously contradicts the Causality Principle. A reasonable guess for the lower value of τ is, therefore,

$$\tau \sim -\frac{R}{v_i} - \frac{R}{v_f} \tag{22}$$

where R is some mean radius of the repulsive core between A and B.

A positive τ means that when A and B collide they form a state (AB) which is long lived. For such a state there is nothing in classical theory which sets a bound to the lifetime, and therefore, there is no upper limit to the time delay. In addition to this difference between the negative and positive τ values there is a difference in the information content which the relevant trajectories carry. The negative τ trajectories carry essentially information only about the repulsive core. Also, such trajectories are relatively easy to analyze because they describe only one or a small number of collisions between atoms before the system breaks up into C and D. On the other hand, positive time-delay trajectories may explore the whole potential, and each atom in a molecule may have enough time to be affected by the presence of all other atoms in (AB). Therefore, the properties of such trajectories depend more or less on all parts of the potential, and as a consequence their analysis is quite complicated.

In a molecule-molecule collision experiment we cannot follow a single trajectory. We can only measure a cross-section σ, which is averaged over an ensemble of trajectories. As we have seen earlier, each trajectory from the ensemble connects initial and final observable states, which differ from one another in the nonobservable initial conditions. However, the trajectories also differ in their time delays, though we cannot exclude the accidental possibility that some trajectories have the same τ. Since time delays can be measured, at least in principle, we can also measure the contribution of a particular trajectory to the cross-section. For example, the integral number of scattered molecules in the interval from the moment of collision until a certain time t_0 after collision is proportional to

$$\sigma_{\omega_i \to \omega_f}(\tau) = \frac{1}{\sin\theta} \sum_{\leq t_0} \frac{b}{\left| \dfrac{d\omega_f}{dq_i}, \dfrac{d\theta}{db} \right|} \tag{23}$$

where the sum extends only over the trajectories which have time delays less than or equal to t_0. Therefore, $\sigma_{\omega_i \to \omega_f}$ is now a function of time, as indicated in Equation 23. Such a cross-section provides more detailed information about the system, because with the aid of the time delay we can select trajectories which are otherwise indistinguishable and observe their contributions to the cross-section.

The long-lived trajectories are very difficult to analyze. In particular, it is not simple to answer the question of what causes their longevity. One of the reasons for this difficulty is that the trajectory Equations 16 are nonlinear and therefore many of the standard techniques for analyzing the properties of solutions of differential equations either do not apply or are not good enough. For example, standard perturbation techniques cannot be applied to Newton's equations, except in rather special cases. For the same reason, the analytic structure of the solutions is difficult to obtain, and in particular we do not know the structure of the relationship $\Omega_f\ (Q_i)$, which is central to calculation of the cross-section. We also do not know how τ depends on Q_i. Practically the only thing which is left is the numerical solution of the set of Equations 16 for a particular system. From many numerical calculations we can try to extract some general features which are common to all systems. If possible, we try to understand these features by going back to Newton's equations and see what causes them.

Another approach to the analysis of long-lived states is to use intuition. The most common assumption is that for very long-lived trajectories the distribution $\Omega_f\ (Q_i)$ is random. There is no justification for this except that our common sense tells us this must be true. However, throughout this book we will try to avoid such seemingly plausible assumptions because they may sometimes be deceiving, as will be shown later. On the other hand, very few alternatives are left. One of them is to study what happens to two trajectories which differ slightly in their initial conditions, rather than a single trajectory. In this way we get more understanding about the long-lived trajectories and the relationship $\Omega_f\ (Q_i)$.

In order to shorten the notation we will designate the column matrix of all particle coordinates by r. Likewise $\mathring{r}$ stands for the velocity column vector of all particles. In this notation the equations of motion (Equation 16) are

$$\omega\, \mathring{\mathring{r}} = -\nabla V(r) \tag{24}$$

where ∇ is the gradient operator. Let us assume that two trajectories are initially close together, i.e., their initial conditions are similar. If we designate by ϵ the difference between these two trajectories then the equations of motion for ϵ are obtained from Equation 24

$$\omega\, \mathring{\mathring{r}} + \omega\, \mathring{\mathring{\epsilon}} = -\nabla V(r + \epsilon) \tag{25}$$

and if it is assumed that ϵ is small then[1]

$$\omega\, \mathring{\mathring{\epsilon}} = -\nabla(\nabla V \cdot \epsilon) \tag{26}$$

If we define the matrix

$$V^{(2)}_{i,j} = \frac{\partial^2 V}{\partial r_i \partial r_j} \tag{27}$$

then the approximate solution for ϵ of the Equations 26 is

$$\epsilon = \cos\{[\omega^{-1} V^{(2)}]^{1/2} t\} \cdot \epsilon_0 + [\omega^{-1} V^{(2)}]^{-1/2} \cdot \sin\{[\omega^{-1} V^{(2)}]^{1/2} t\} \cdot \mathring{\epsilon}_0 \tag{28}$$

where ϵ_0 and $\overset{\circ}{\epsilon}_o$ are the initial values of ϵ and $\overset{\circ}{\epsilon}$, respectively. It was assumed that $V^{(2)}$ is independent of time, which means that Equation 28 is valid in the vicinity of t = 0, i.e., locally. By defining a matrix $\omega^{-1/2}$ U, where U is unitary, which diagonalizes $\omega^{-1} V^{(2)}$, then

$$\epsilon = \omega^{-1} U \cos(v^{1/2} t) \tilde{U} \omega^{1/2} \epsilon_0 + v^{-1/2} \sin(v^{1/2} t) \tilde{U} \omega^{1/2} \overset{\circ}{\epsilon}_0 \tag{29}$$

where v is diagonal. We notice that as long as all elements of v are positive, ϵ is oscillatory in time. The same conclusion is valid globally, i.e., along the entire length of trajectory, if all the eigenvalues v are positive for all t. If this is true, for any two close trajectories, then the system is called stable. Therefore, in a stable system any two initially close trajectories will remain so along their whole length. In particular, this is true if the system consists of coupled harmonic oscillators, in which case $V^{(2)}$ is constant and the solution of Equation 28 is valid for all times. Two close trajectories in such a system remain so at all times. However, if some of the oscillators are replaced by a parabolic potential with negative sign (i.e., by the repulsive linear force) then some of the eigenvalues v can be negative. In such a case ϵ increases exponentially with time. This conclusion is valid more generally, i.e., if some eigenvalues v of the r-dependent matrix $\omega^{-1} V^{(2)}$ are locally negative, then the separation between two neighboring trajectories increases exponentially in time.[2] Such systems are called unstable.

In general, trajectories go through regions of stability and instability and are seldom entirely of one kind. This means that two close trajectories always diverge for a general potential, but the rate depends both on how long a time they spend in the unstable region and how negative some of the eigenvalues v are. It is obvious that the longer the trajectory spends in a region of instability, the more it will increase its separation from a neighboring one. The immediate consequence for the unstable systems and long-lived trajectories is that the canonical variables Ω_f are rapidly varying functions of the initial conditions Q_i. This we can demonstrate through the solution of Equation 24. Initially, Ω_i are fixed, by assumption, and Q_i have the values which take the trajectory to the final canonical variables Ω_f. If we designate the change in the nth variable $(Q_i)_n$ by $\delta(Q_i)_n$ then the coordinate r_m will change by

$$\delta r_m = \frac{\partial r_m}{\partial (Q_i)_n} \delta(Q_i)_n \tag{30}$$

and similarly for $\overset{\circ}{r}_m$. The change of δr_m in time is given by the solution of Equation 24, and since $\delta(Q_i)_n$ is common to all components of r and $\overset{\circ}{r}$, we have approximately

$$\frac{\partial r}{\partial (Q_i)_n} = \omega^{-1} U \Big\{ \cos(v^{1/2} t) \tilde{U} \omega^{1/2} \left[\frac{\partial r}{\partial (Q_i)_n}\right]_0 + v^{-1/2} \sin(v^{1/2} t) \tilde{U} \omega^{1/2} \left[\frac{\partial \overset{\circ}{r}}{\partial (Q_i)_n}\right]_0 \Big\} \tag{31}$$

and similarly for $\frac{\partial \overset{\circ}{r}}{\partial (Q_i)_n}$. Of course, the solution of Equation 31 is not valid over a long span of time, but if some eigenvalues v are negative along most of trajectory, then we can write an estimate

$$\frac{\partial r_m}{\partial (Q_i)_n} \sim 0(e^{ct}) \tag{32}$$

where c is some positive constant which is, loosely speaking, some average value of $|v^{1/2}|$ of the most negative v along the whole trajectory. The appropriate change in the lth canonical variable $(\Omega_f)_l$ is

$$\delta(\Omega_f)_l = \sum_m \frac{\partial(\Omega_f)_l}{\partial r_m} \frac{\partial r_m}{\partial(Q_i)_n} \delta(Q_i)_n \tag{33}$$

and therefore we have an estimate

$$\frac{\partial(\Omega_f)_l}{\partial(Q_i)_n} \sim 0(e^{ct}) \tag{34}$$

which indeed shows that for the predominantly unstable systems the rate of change of the final canonical variables, with respect to Q_i, is very rapid. From this we also obtain an estimate of the cross-section

$$\sigma_{\Omega_i \rightarrow \Omega_f} \sim 0(e^{-ct}) \tag{35}$$

The expression in Equation 35 describes the qualitative behavior of the time dependence of the cross-section for unstable systems and long-lived trajectories. It predicts a rapid fall-off of the cross-section with time.

By the same arguments we can show that for stable systems, or those which are predominantly stable, the rate of change of Ω_f with Q_i is independent or nearly independent of time. This also means that the cross-section can be approximated by

$$\sigma_{\Omega_i \rightarrow \Omega_f} \sim 0(t^0) \tag{36}$$

meaning that it is constant, or nearly constant, in time.

The estimate of the separation rate of two close trajectories has also more practical consequences. Usually the trajectories are obtained by numerical integration of the set of Equations 16, and for this purpose various numerical algorithms are used. These algorithms are designed to propagate a solution by a single step Δ, and at the end of it, the result is an approximation to the exact solution. The approximation is measured by the power of Δ and therefore the nth order algorithm produces a solution which deviates from the exact one by Δ^{n+1}. This is so-called local error. The initial conditions for the next step deviate from those which would have been obtained from the exact solution by an error of the order Δ^{n+1}. Therefore, the algorithm will now propagate a trajectory with slightly different initial conditions. These inaccuracies are repeated at each step and accumulate along the trajectory. After k steps the approximate trajectory will deviate from the exact one, but this error may considerably deviate from $k\Delta^{n+1}$, especially for the unstable systems and long-lived trajectories. This can be demonstrated if we assume that after the first step the trajectory is integrated exactly, i.e., by an algorithm of infinite order. In such a case local error would suggest that at the end of integration the deviation from the exact trajectory is of the order Δ^{n+1}, though this is not true for the unstable systems. We have shown that for such systems the separation from the exact trajectory is of the order $0(e^{ct})$, and in our example, it will be $\Delta^{n+1}\ 0(e^{ct})$, which for a long-lived trajectory may be quite large. This is an important conclusion which points to some very serious difficulties when integrating over long-lived trajectories. Although the local error in integrating over such trajectories can be quite small, the solution can deviate considerably from the exact one. Therefore, controlling only the local error at each step (e.g., by checking the total energy, which must never deviate from the constant value by more than Δ^{n+1}) is not sufficient to ensure an accurate trajectory because any small local error in the beginning causes the integration to follow a solution which has different initial conditions from those of the exact trajectory. In other words, controlling the local error for unstable systems is not sufficient for controlling the global error of the solution. Controlling the global error is, in fact, much more difficult than controlling the local error. The only reliable way of checking it is by the time-reversed integration of trajectory, i.e., at the end of trajectory the integration is continued backward in time until the initial point is reached. The deviation from the initial conditions so obtained is a measure of the global error.

As we have said, there is not much that we can get from general consideration of the Equations 16. We can, however, get some additional insight into the long-lived states by using models which reproduce, more or less, the essential features of the systems and at the same time are simple enough to be solved, hopefully, analytically. One such model,[3] is based on the assumption that forces among atoms are harmonic up to certain interatom separations, beyond which the forces are zero. In other words, molecules are treated as a set of coupled harmonic oscillators. Therefore, the potential for a pair of atoms i and j can be written

$$V_{i,j} = \begin{cases} \omega_{ij}(\mathbf{r}_i - \mathbf{r}_j - \boldsymbol{\delta}_{ij})^2 + \epsilon_{ij}; & |\mathbf{r}_i - \mathbf{r}_j| \leq \Delta_{ij} \\ 0; & |\mathbf{r}_i - \mathbf{r}_j| > \Delta_{ij} \end{cases} \tag{36a}$$

where $\boldsymbol{\delta}_{ij}$ is the equilibrium vector of the two atoms, Δ_{ij} is some critical interatom separation, and ϵ_{ij} are energies which ensure that V_{ij} is continuous. The overall potential is a sum of all pair potentials V_{ij}, i.e.,

$$V = \sum_{i>j} V_{ij} \tag{37}$$

This potential describes two essential features of molecules; their vibrations and dissociations. Moreover, the appropriate equations of motion (Equations 16) can be solved analytically, which makes this model very attractive for studying molecular dynamics. In fact, Slater used this model for just this reason; to predict the rates of unimolecular reactions. Although today there are much more reliable statistical models, the Slater model is very attractive because it is so simple.

The potential (Equation 37) is in general a bilinear form, which can be written as

$$V_x = V_{xo} + \sum_{i \neq j} V_{ij}\, e_i\, e_f \equiv V_{xo} + \tilde{e}\, v\, e \tag{38}$$

where e_i are the relative x-coordinates and V_{xo} is a constant. Similarly, the bilinear form for the y and z relative coordinates can be written. The Lagrange function for the motion along the x-axis is

$$S_x = \frac{1}{2} \mathring{\tilde{e}}\, \tilde{A}^{-1}\, m\, A^{-1}\, \mathring{e} - V_{xo} - \tilde{e}\, v\, e \tag{39}$$

which is a combination of two bilinear forms. These bilinear forms can be diagonalized simultaneously by the transformation

$$e = W\, \omega^{-1/2}\, T\, u \tag{40}$$

where W and ω were defined in Equation 12. The matrix T which is unitary will be defined shortly. In the new coordinates u, usually called the "normal coordinates", the Lagrange function S_x is

$$S_x = \frac{1}{2}\mathring{\tilde{u}}\mathring{u} - V_{x0} - \tilde{u}\tilde{T}\omega^{-1/2}\tilde{W}vW\omega^{-1/2}Tu \qquad (41)$$

and if we choose T such that

$$\tilde{T}\omega^{-1/2}\tilde{W}vW\omega^{-1/2}T = \nu^2 \qquad (42)$$

where ν^2 is diagonal, then the equations of motion for u are

$$\mathring{\mathring{u}}_i = -\nu_i^2 u_i \qquad (43)$$

which have the general solution

$$u_i = A_i \cos(\nu_i t + B_i) \qquad (44)$$

where A_i and B_i are determined by the initial conditions. Therefore, the coordinate q_i has a general time dependence

$$q_i = \Sigma\, a_{ij} \cos(\nu_j t + B_j) \qquad (45)$$

Similarly, we obtain the equations of motion along the y- and z-axis.

Dissociation of a bond between the atoms i and j will occur when

$$|\mathbf{r}_i - \mathbf{r}_j| = \Delta_{ij} \qquad (46)$$

and the time when this happens can be determined from the explicit equation of motion.

Although the original application of this model, as we have said, was to predict the dissociation rates for an energized molecule, it can be used to describe collisions between two molecules, but this is more difficult. The main obstacle is technical, because when molecules collide several bonds can be formed in succession, and the exact sequence of their formation can be quite complicated. The collision between an atom and a molecule can, however, be handled in this way as the incoming atom may be treated as a free particle until it reaches the interaction region of one of the atoms in the molecule. At this moment the atom becomes a member of the enlarged molecule, which has its own normal coordinates. The time evolution of the new normal coordinates is given by Equation 45, with initial conditions which are determined from the relative velocities of the atom and molecule at the moment of impact, and the phases and amplitudes of normal coordinates of the molecule at the same instant. Having now the complete solution for the time evolution of the systems, we can follow the changes in the relative separation of atoms in such a molecule until the time when dissociation occurs. At this moment the collision is over, and we can determine the final distribution of atoms and associated energies, i.e., the set Ω_f.

The drawback of this model is that it is representative of a typical stable system, and therefore in that sense it is not very realistic. What this means, in fact, is that the model does not take into account the anharmonic effects which introduce the instability into the system. These effects are nonnegligible for large amplitude atom motion which in turn is connected with high energy motion of atoms in molecules. Therefore, the model can be useful for describing low energy collisions provided that, during most of the lifetime of the long-lived state, the energy of the incoming atom is more or less uniformly distributed among

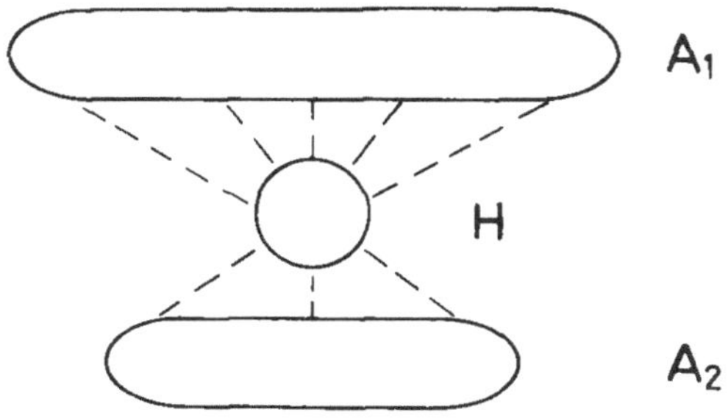

FIGURE 1. Molecule which contains two species separated by a heavy atom H In collision with an atom, the final states of this molecule may not be random.

all atoms. In such a case the vibrations are almost harmonic, except for the short time between the formation of the long-lived state and its dissociation.

There is no simple, exactly solvable model which is typical for the unstable systems, and therefore, one can only guess at their properties. It is believed that because of the exponential separation in time of two initially close trajectories, the final distribution of states is random. In other words, shortly after the long-lived state is formed the trajectory "forgets" its initial conditions, and therefore any final state allowed by the integrals of motion (e.g., conservation of energy and total angular momentum) is *a priori* equally probable. In fact, a single trajectory does not "forget" its initial conditions, because no matter how long it is, it can always be traced back to its initial conditions. What is meant is that because of the exponential separation of trajectories, any small change in the initial conditions of one of them causes such a large change in the final state that their distribution looks random. Therefore, it would be more appropriate to say that for the long-lived states of the unstable systems the final distribution of states is random when initial values of the conjugate variables Q_i are changed. Strictly speaking, even this is not true because for a sufficiently small change of the initial variables Q_i, the final states will also change smoothly, except in the vicinity of certain isolated points, called catastrophies. This is the consequence of two facts; the lifetime of the long-lived states is finite and the gradient of potential is assumed to be a continuous function of coordinates. However, over the range of the values Q_i the final distribution of states may appear to be random.

The assumption of the random distribution of the final states must be made with great caution. Very often the long-lived trajectory cannot "explore" all available states because it does not live long enough to do this. One situation where this may happen is in the case of a molecule in which a heavy atom divides the molecule into two distinct parts, as shown in Figure 1. Here this atom is designated H.

The motion of the atoms in fragment A_1 is almost independent of the motion of the atoms in fragment A_2 if direct interaction between the two fragments is weak. This means that if in the collision with an atom a long-lived state is formed with one of the fragments in Figure 1, its motion and dissociation will be essentially independent of the other fragment. Therefore, the final distribution of states cannot be random, since those which are the mixture of the fragments A_1 and A_2 are missing.

In the previous example we have demonstrated one possibility where the final distribution of states, after dissociation of the long-lived state of an unstable system, is not random. There may be other cases in which the masses of atoms do not play a critical role. It may be just the configuration of the potential which prevents easy access to all final states. In short, the greatest obstacle to finding the random distribution of final states is the finite lifetime of the long-lived states, and exponential separation of trajectories cannot be the only criterion.

In the beginning it was stated that classical theory is an initial value problem, and thus the equations of motion are relatively straightforward to solve, at least numerically. We have discussed several drawbacks of classical theory, and among the most serious ones is the fact that mathematical modeling (e.g., apply the perturbation theory, analyze the analytic properties of solutions, etc.) is difficult. The limitations of the theory are discussed in the following section.

II. LIMITATIONS OF CLASSICAL THEORY

The use of Newton's equations of motion for the description of atom and molecule collisions has some obvious and some less obvious limitations. The obvious limitation is that atoms and molecules are quantum objects and therefore, strictly speaking, the use of the classical equations of motion is not permitted. However, their use for the calculation of cross-sections in quantum systems is not recent. Classical theory was used very early by Rutherford[4] to predict the cross-sections for collisions of α-particles with nuclei under the influence of Coulombic forces. This is, of course, a quantum system, but at that time there was little awareness of this. For this system both the classical and quantum cross-sections are exactly equal which is not entirely accidental. It would take us outside the main scope of this book to discuss the reason why such equalities happen, but as a rule of thumb, one can say that whenever initial and final states are connected by only one trajectory with a small time delay, the classical and quantum cross-sections are equal. For example, in the case of high energy elastic scattering of atoms, when orbiting does not occur, only one trajectory leads from the initial point to a large scattering angle. As a consequence the two theories yield exactly equal results for the large angle atom-atom elastic differential cross-section.[5,6] However, these cases are of no particular interest to us because for such scattering long-lived states are not involved.

However, whenever several trajectories connect an initial and final state, quantum and classical theories predict different cross-sections. Since in the quantum theory there is no concept of a trajectory it is not easy to see the reason for this. In the Feynman and Hibbs[7] formulation of quantum mechanics the reason becomes more obvious. In this formulation the initial and final states are connected by an infinite number of trajectories. Only some of them obey the classical equations of motion and are called the classical trajectories. With each trajectory is associated an amplitude, in an analogous way to that of the partial cross-section in classical theory. These amplitudes are in general complex numbers, and the module squared of their sum is the quantum cross-section. The difference between such a cross-section and the classical one is not just in the fact that nonclassical trajectories are taken into account, but also in the way in which their contributions to the cross-section are taken into account. If we write the amplitude associated with i^{th} trajectory (for simplicity, only the classical one will be taken into account) as f_i, then the appropriate cross-section is

$$\sigma = \left|\sum_i f_i\right|^2 = \sum_i |f_i|^2 + 2\sum_{i>j} \mathrm{Re}(f_i^* f_j) \tag{47}$$

which formally differs from Equation 20 in the cross-products between different amplitudes. This term can be significant, and therefore the classical and quantum cross-sections can sometimes be quite different.

In the case of inelastic collisions involving long-lived states, the classical theory of scattering may have additional difficulties. Measurement of the time-delay requires accurate determination of the initial and final position and velocity of the molecules, otherwise this quantity loses its meaning. However, we have said earlier that there is a limitation to the accuracy with whicn this measurement can be achieved, given by the uncertainty principle

(Equation 17). Therefore, if the collision energy of the two molecules is known exactly, the initial relative position of molecules is completely undetermined, and therefore, even in principle, we would not be able to measure the time delay. Likewise, if the initial relative positions of molecules are known exactly, the collision energy is undetermined, and again we cannot determine the time delay. The optimum situation is one in which there is some uncertainty in both variables in which case the time-delay is determined with a certain error and its value is in direct relation to the masses of the molecules, as follows from Equation 17. However, within the error tolerance there is a whole set of trajectories with slightly different initial conditions connecting initial and final states. For the unstable systems all of the trajectories separate exponentially, but since they connect the same initial and final states, their lifetime can be very different. In order to see this more clearly, let us consider two trajectories from this set which have slightly different initial conditions for the relative separation and velocity.

If these two trajectories have the same initial variable q_i conjugate to the initial canonical variable w_i, then for long-lived states and unstable systems the final state w_f would be very different, which follows from our earlier discussion about these systems. However, if the final values of w_f are fixed for two trajectories, as is required by experiment, then their initial values q_i must be different, and this difference can be large. Therefore, two trajectories which connect the same initial w_i and final w_f have different initial conditions q_i, and in general they will have quite different time delays.

This reasoning applies to all trajectories which have initial conditions within an interval determined by the uncertainty principle, and all will contribute to the cross-section. Hence the true cross-section (Equation 23) is modified by the contributions of such trajectories. *A priori* it is not possible to say what temporal part of the cross-section (Equation 23) will be affected most, because the trajectories may have different lifetimes. As a consequence much vital information about the system can be lost as the result of the uncertainty principle. It should be pointed out that when a system is stable, i.e., trajectories do not separate exponentially, then the uncertainty in the initial conditions affects the cross-section linearly.

There is another limitation of classical theory; it does not describe resonances, i.e., the typically long-lived quantum states. Although, as we shall see later, resonances are not long-lived states in a true sense, they nevertheless are very often treated as if they were. Therefore, our discussion of resonances in the content of the classical theory must be carried on with caution. With this proviso always considered, we shall use classical theory to reproduce some of the properties of resonances using a model that assumes they are long-lived states. A detailed discussion of what a resonance is will be given in the following chapter. In short, it is an effect which has its origin in the wave character of particles and is usually associated with the presence of standing waves. There are no general rules which govern the way such standing waves arise. Their appearance depends on the circumstances; sometimes a wave is trapped in a region defined by the shape of a potential barrier (in much the same way as a water wave). In this case a standing wave is formed for a particular collision energy. The wave gets inside this region by tunneling. In other circumstances, when a colliding atom forms a compound state with a target molecule, a resonance is observed if the energy coincides with that of a quantized energy level. Broadly speaking, this state is also a multidimensional standing wave. There are other possibilities for the formation of resonances, but these are two typical cases. In general, however, for complex systems it is more difficult to trace the origin of a resonance, and therefore their study is usually limited to very simple systems, e.g., atom-atom or atom-diatom collisions.

It is characteristic of resonances that they occur at particular collision energies, the ones which correspond to the formation of standing waves. Their formation is recognized as rapid changes of the cross-section in the small intervals around these energies. More about their properties will be presented later, but here we shall examine the ways in which the classical

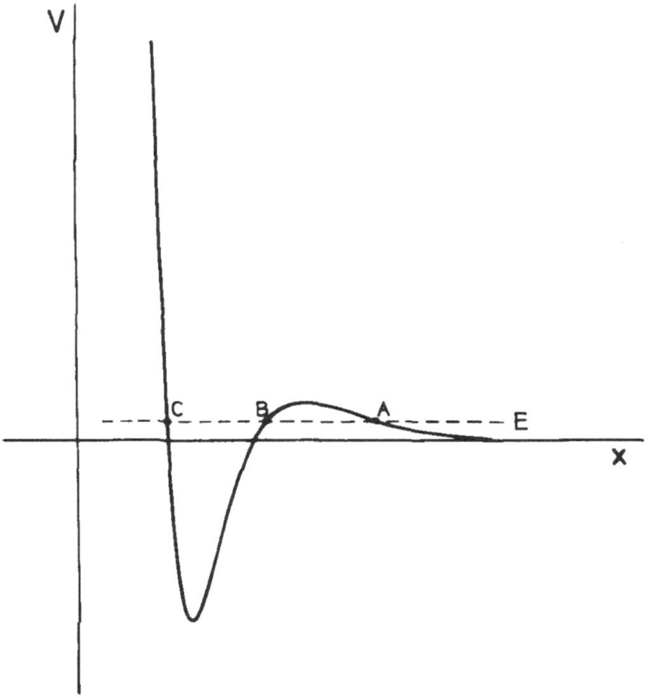

FIGURE 2. Atom-atom potential with a barrier. Points A, B, and C indicate where kinetic energy of relative motion of atoms is zero

description of resonances is deficient. We will take two typical cases of resonance phenomena. In the first case we shall consider an atom-atom collision governed by a typical one-dimensional potential with a barrier, shown in Figure 2. It is assumed that the total collision energy E is below the top of the barrier, in which case there are three points, indicated by A, B, and C in Figure 2, where the kinetic energy of the particle is zero. These three points also divide the physical from the nonphysical regions of the potential, and the criterion for this is that the kinetic energy of a particle is positive or negative, respectively.

The motion of a particle, if the units are conveniently chosen, is described by the equation

$$\frac{1}{2}\ddot{x} = -\frac{\partial V}{\partial x} \tag{48}$$

The particle comes from the right, and far away from the origin it has velocity v_o. As it approaches A, it slows down, halts exactly at A, turns, and journeys back towards $x = \infty$. The motion of the particle is predicted without, in fact, solving the equation of motion (Equation 48). The main physical argument adhered to is that the particle cannot reach points to the left of A because this would necessitate negative kinetic energy which is classically impossible because this would require the velocity of particle to be imaginary. Such quantities are meaningless in the real world, and therefore the particle is restricted to the region to the right of A. However, if we relax this requirement by allowing the *time* variable to take on complex values, then the particle can "tunnel" through the space between A and B to reach the region between B and C, where again its kinetic energy is positive. There are two formal arguments which allow us to do this. In one, Equation 48 can be formally analyzed as a function of one variable (which has certain analytic properties) by extending the definition of the time variable to possess real and complex values, so obtaining the necessary analytical continuation of the solutions. The other argument is based on quantum arguments. Since a trajectory has no meaning in quantum systems, we can assume that it is complex in the interaction region. The only requirement is that for that part of the trajectory outside the interaction region the time variable must be real. In other words, the initial motion of the particle is in real time, in the interaction region it may be in complex time, but when it comes out it must again be in real time. The last requirement is purely physical. The time delay is observable and therefore it must be real. Since this quantity is essentially the time interval during which interaction occurs, it is obvious that after collision time must be real.

We will now show how such analytic continuation is possible and that some, but not all, properties of resonances can be reproduced. Let us assume that at time $t = 0$ the particle is at A. To the left its velocity is imaginary, but if the time becomes imaginary then the coordinate x is real. We may choose the sign of the imaginary velocity, but if the sign of the imaginary time is the same, then the coordinate x decreases towards the point B. The time required to get from A to B is obtained from the conservation of energy as

$$t_{AB} = \frac{1}{2}\int_{x_A}^{x_B} \frac{dx}{(E - V)^{1/2}} = -i\int_{x_A}^{x_B} \frac{dx}{2(V - E)^{1/2}} \tag{49}$$

where $V - E > 0$. At this time t_{AB} the particle is at B and then continues to travel to the left with real velocity, in which case the time increment is now also real. In the time interval

$$t_{BCB} = \int_{x_C}^{x_B} \frac{dx}{(E - V)^{1/2}} \tag{50}$$

the particle travels from B to C and back again to B. There is a choice now for the particle either to return to C or to tunnel to A. If the time continues to increase along the real axis, it will return to C, but if it becomes imaginary it will tunnel through the barrier to A. Let us assume that the time becomes imaginary with opposite sign to t_{AB}. The sign of the imaginary velocity must now be the same as that of t_{AB}, and therefore when the particle returns to A it has spent a total time $t = t_{AB} + t_{BCB} + t_{BA}$ traveling in the region $x < x_A$.

The time intervals t_{AB} and t_{BA} are imaginary and of opposite sign; therefore, the time interval $t_{BCB} > 0$. This means that for an observer who "sends" particles from far away it would appear that there are two kind of particles: those which return with a time delay corresponding to reflection at the point A and those which return with a longer time delay. The time lag between these two is t_{BCB}. Of course, such an observation assumes that the particle did not make an internal reflection at point B. If it did, then there would be a third signal lagging behind the first one by $2 \times t_{BCB}$. In fact, there would be a progression of signals with the time interval t_{BCB} between them that would arise due to multiple reflection at B.

Once the particle finds itself between the turning points B and C, it is in fact bounded. This means that its energy can have only discrete values, given by the Bohr-Sommerfeld rule

$$\int_{x_0}^{x_B} (E - V)^{1/2}\, dx = (n + 1/2)\, h \tag{51}$$

This restriction is still in accordance with classical dynamics since the concept of a trajectory is preserved. The meaning of Equation 51 is that only trajectories of certain energy can tunnel from A to B and in all other cases they are reflected at A. For the observer the outcome of his experiment would be now slightly different. In general, he would observe only particles with one time delay, corresponding to reflection at A, except for some discrete values of the collision energy when particles with longer time delays would be observed. He could say that owing to a resonance, a longer time delay is observed.

Of course, an experiment with classical particles can have only one outcome; all particles are reflected at A. However, the point of our discussion was to show that within classical theory it is possible to predict the formation of a resonance. For this purpose we required two assumptions: (1) the classical equations of motion also have meaningful solutions in the complex time plane and (2) the periodic orbits (i.e., those between the points B and C) are quantized. The prediction of resonances does not, however, mean that classical theory can necessarily describe all their properties. For example, their lifetimes and their widths are not determined by classical theory. Classical theory also cannot predict resonances which are above the barrier unless the Bohr-Sommerfeld rule is generalized to the potentials with complex turning points. Classical theory can also not handle collisions when states are electronically excited. Strictly speaking, solving this problem in classical theory means setting up Newton's equations for all electrons and nuclei and following the time evolution of such a system. It is known, however, that classical theory is very poor in describing the motion of electrons; in particular it does not take into account the Pauli exclusion principle, and it cannot describe the exchange properties of electrons which have an important role in molecular interactions. On the other hand, if the Bohr-Oppenheimer separation is made first and then the classical equations are applied to the nuclear motion, problems arise in the treatment of the multivalued potentials and the cross-terms among them. In the simplest case this problem is treated with two electronic states, one in the ground state and the other in some excited state with a different potential.

If the motions of atoms can be described by diabatic potentials, then for some interatom distance the two potentials may cross. At such a crossing point classical theory, without any additional assumptions, is unable to predict which potential curve the particle will follow. It can be assumed, as in the case of tunneling, that the trajectory "splits" and part of it follows one potential, the other part follows the other potential, and both contributions are added in the cross-section. However, there is no way, in strictly classical theory, to decide about the relative weights of these two parts of trajectory, nor can the classical theory solve the problem of the collision between two atoms governed by adiabatic potentials. It becomes even more complicated because now the two potentials do not cross for real interatom distance. They may cross in the complex coordinate plane which means that (as in the case of tunneling) trajectories must become complex if one wants to take into account transitions from one electronic state to the another.[8]

In the above discussion we have demonstrated the limitations of the classical theory of atom and molecule collisions, with the emphasis on the description of long-lived states. The discussion is not complete, but the major points have been covered. Some of the limitations can be overcome by semiclassical theory, but some are inherent in the concept of the classical trajectory and can only be properly treated by the quantum theory of collisions.

Chapter 2

QUANTUM THEORY OF SCATTERING

I. GENERAL ASPECTS

The basis of the quantum theory of scattering is the time-dependent Schrödinger equation

$$i\hbar \frac{\partial \psi}{\partial t} = H \psi \tag{52}$$

where H is the Hamilton operator

$$H = -\sum_{j=1}^{n} \frac{\hbar^2}{2m_j} \left(\frac{\partial^2}{\partial x_j^2} + \frac{\partial^2}{\partial y_j^2} + \frac{\partial^2}{\partial z_j^2} \right) + V \tag{53}$$

The sum in Equation 53 extends over all particles in the system, and V is the potential among them. The mass of the jth particle is m_j.

Mathematically speaking quantum theory is both an initial *and* boundary value problem because the solution of the Schrödinger equation (Equation 52) has certain properties for large interparticle separations, and it evolves in time from a certain initial value. In addition it is the theory of linear partial differential equations, and therefore many properties of the wave function can be obtained by the powerful techniques of mathematical physics. Some of these techniques will be introduced along with the discussion of the long-lived states in quantum systems.

The basis of all applications of the Schrödinger equation lies in the physical interpretation of its solution ψ, the wave function. Its square modulus give the probability density of finding the system at points with coordinates (x_j, y_j, z_j); $j = 1, 2, \ldots, n$ at time t. Therefore, the probability of finding the system in a volume element $dV_1\, dV_2 \ldots dV_n$ and in the time interval dt is

$$dP = |\psi|^2\, dV_1\, dV_2 \cdots dV_n\, dt \tag{54}$$

where $dV_j = dx_j\, dt_j$. In general, the wave function is complex and ψ^* satisfies the equation

$$-i\hbar \frac{\partial \psi^*}{\partial t} = H \psi^* \tag{55}$$

If Equations 52 and 55 are multiplied from the left by ψ^* and ψ, respectively, and then subtracted from each other, an important relationship is obtained

$$i\hbar \frac{\partial}{\partial t} (\psi\psi^*) = -\frac{\hbar^2}{2} \sum_j \frac{1}{m_j} \nabla_j(\psi^* \nabla_j \psi - \psi \nabla_j \psi^*) \tag{56}$$

Where ∇_j is the gradient operator in the jth coordinate. It represents the conservation of the probability equation in the differential form or the continuity equation. By analogy with the continuity equation in electrodynamics, $\psi\psi^*$ plays the role of the charge density and

$$\mathbf{J}_j = \frac{\hbar}{2im_j} (\psi^* \nabla_j \psi - \psi \nabla_j \psi^*) \tag{57}$$

is the current density.

The Hamiltonian equation (Equation 53) is given in the L system of coordinates which can be written in a matrix form

$$H = -\frac{1}{2}\tilde{\nabla}\, m^{-1}\nabla + V \tag{58}$$

where we have used the units for the energy in which $\hbar = 1$. The operator ∇ is a column vector and m^{-1} is diagonal matrix. Transformation of Equation 58 into the C system of coordinates is not unique, because it all depends on how the relative coordinates of the atoms are defined. We usually define them for a given convenient situation. For example, if two molecules A and B, with respective number of atoms n_A and n_B, are involved in a collision, then the potential V has the form

$$V = V_A + V_B \tag{59}$$

when they are far away from each other. The part V_A contains only the coordinates of the atoms in A, and V_B contains only the coordinates of atoms in B. It is only possible to separate the potential V as in Equation 59, where each part is a function of its own group of coordinates, with aid of relative coordinates which have certain specific properties. The atoms should be conveniently numbered and then the coordinates as defined in Equation 6 used.

In classical theory we did not pay much attention to this inadequacy of the coordinates (Equation 6) because the numerical solution of Equation 16 is quite straightforward in any coordinates. However, quantum theory is also a boundary value problem, and it is quite important to choose coordinates which reflect the fact that molecules A and B are independent. The way to do this is to define relative coordinates (Equation 6) for each molecule separately so that the kinetic energy part of the Hamiltonian has the form

$$T = T_A + T_B - \frac{1}{2\Sigma_A}\Delta_A - \frac{1}{2\Sigma_B}\Delta_B \tag{60}$$

where Σ_A and Σ_B are the total masses of the molecules A and B, respectively, while the kinetic energy operator for A is

$$\Delta_A = \frac{\partial^2}{\partial R_{xA}^2} + \frac{\partial^2}{\partial R_{yA}^2} + \frac{\partial^2}{\partial R_{zA}^2} \tag{61}$$

and similarly for B. The set of coordinates (R_{xA}, R_{yA}, R_{zA}) gives the position of the center of mass of A. The kinetic energy operator T_A for the relative coordinates is

$$T_A = -\frac{1}{2}\sum_{j=1}^{n_A-1}\frac{m_j + m_{j+1}}{m_j\, m_{j+1}}\frac{\partial^2}{\partial q_j^2} + \sum_{j=2}^{n_A-1}\frac{1}{m_j}\frac{\partial^2}{\partial q_{j-1}\,\partial q_j} \tag{62}$$

where m_j; $j = 1, 2, \ldots, n_A$ are the masses of atoms in the molecule A.

The kinetic energy T can be transformed further if we define the relative coordinates $Q = (Q_x, Q_y, Q_z) = (R_{xA} - R_{xB}, R_{yA} - R_{yB}, R_{zA} - R_{zB})$ and the coordinates $R = (R_x, R_y, R_z)$ of the center of mass of both molecules. In such a case T is

$$T = T_A + T_B - \frac{1}{2\,\mu_{AB}}\Delta_Q - \frac{1}{2\,\Sigma}\Delta_R \tag{63}$$

where $\mu_{AB} = \Sigma_A\Sigma_B/\Sigma$ and $\Sigma = \Sigma_A + \Sigma_B$. In these coordinates as $Q \rightarrow \infty$ the potential V separates into two parts, as in Equation 59, and therefore the Schrödinger equation becomes

$$i\hbar \frac{\partial \psi}{\partial t} = \left(H_A + H_B - \frac{1}{2\,\mu_{AB}}\Delta_Q\right)\psi \tag{64}$$

where we have omitted the operator which corresponds to the motion of the center of mass of the whole system. When the two molecules are far apart it can be assumed that they are in a particular eigenstate so that the wave function ψ can be written as a product

$$\psi = \psi_A\psi_B\varphi \tag{65}$$

where ψ_A and ψ_B are the eigenfunctions of H_A and H_B, respectively. The wave function φ corresponds to the relative motion of the two molecules and satisfies the equation

$$i\hbar \frac{\partial \varphi}{\partial t} = \left(E_A + E_B - \frac{1}{2\mu_{AB}}\Delta_Q\right)\varphi \tag{66}$$

The value of φ is determined by the initial condition at $t = 0$, which we designate by φ_0, and this in turn is determined by experimental circumstances. Let us assume that in the experiment the initial relative motion of the two molecules is along the Q_z-axis. Let us also assume that at $t = 0$ the uncertainty in the relative separation of the two molecules is small. The reason for this is that if the time delay is to be measured then it is necessary to determine where the two molecules were at $t = 0$. If this is not known it would not be possible to determine accurately the time and place of arrival of molecules after collision. On the other hand, the uncertainty in the relative position cannot be small because the uncertainty in the initial relative velocity of the two molecules would then be large. Therefore, some optimal intermediate uncertainty is required. It should be noted that in the usual approach to scattering, the uncertainty in Q_z is large because information about the time delay is not essential.

The uncertainty perpendicular to the Q_z-axis can be large because this error does not significantly perturb the time delay measurement. Therefore, it can be assumed that φ_0, usually called the wave packet, is independent of Q_x and Q_y and $|\varphi_0|^2$ has a certain width around the initial separation Q_0. If the average initial relative momentum of the two molecules is k^0_{AB} the φ_0 can be written in the form

$$\varphi_0 = e^{ik^0_{AB}Q_z}\,\omega_0 \tag{67}$$

where ω_0 is real with a maximum for $Q_z = Q_0$. A typical initial wave packet is shown in Figure 3a, where the interaction region is represented by the shaded region. The contours connect points of equal probability for finding the molecules.

The propagation of the wave function φ in time is given by

$$\psi = \int d^3k\, e^{-iEt}\,\varphi(\mathbf{k};\, \mathbf{q}_A,\, \mathbf{q}_B,\, \mathbf{Q})\, A(\mathbf{k}) \tag{68}$$

where $\mathbf{k}$ is the relative momentum of the molecules A and B. The integration is over the entire volume element in momentum space k. The initial condition for φ at $t = 0$ determines the function A(k) from the integral

$$\psi_A\psi_B\varphi_0 = \int d^3k\, \varphi\, A \tag{69}$$

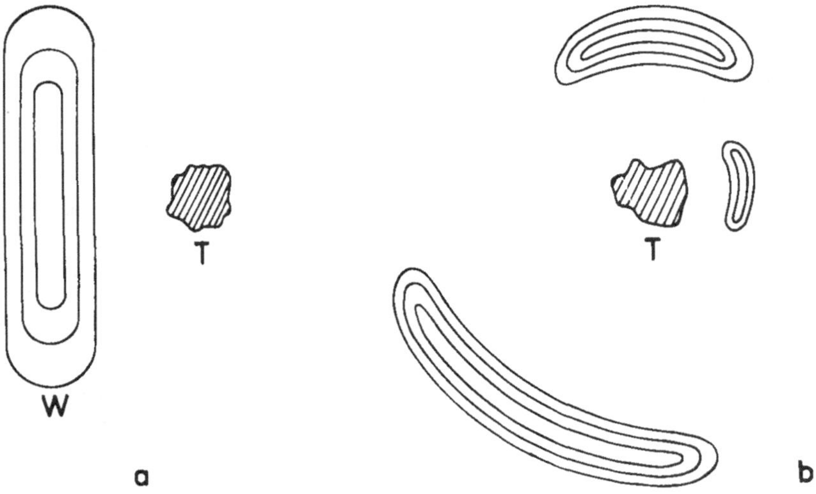

FIGURE 3 (a) Initial location of wave packet W relative to target T. Lines of equal probability in W are shown (b) Scattered wave packet

which is obtained if the exact form of φ is known. This function must be a solution of the time-independent Schrödinger equation

$$H\,\varphi = E\,\varphi \tag{70}$$

where H is the total Hamiltonian in the C system. Among all possible solutions for φ only those with the proper boundary conditions must be chosen. For scattering problems this condition on φ is determined for large separation of molecules, and it is

$$\varphi \sim \psi_A\psi_B\, e^{ik_{AB}Q_z} + \sum_{CD} \frac{1}{Q_{CD}}\, e^{ik_{CD}Q_{CD}}\, f_{AB\to CD}\, \psi_{CD} \tag{71}$$

where CD designates one of the possible reaction channels, together with the appropriate quantum numbers. ψ_{CD} is the eigenfunction for the molecules in the channel CD, $f_{AB\to CD}$ is the scattering amplitude, and Q_{CD} designates the modulus of the separation of the molecules in the channel CD (it is assumed for simplicity that only two molecules are in the final states).

At this point we should notice an important difference between the classical and quantum theories of scattering. In the first, there is no need for prior knowledge of the outcome of a collision in order to obtain the classical trajectory as it is completely determined by the initial conditions. On the other hand, in the quantum theory the knowledge of outcome determines the time evolution of ψ, which is a serious drawback in the theory. This drawback is explicitly evident in the form of φ; the system of coordinates changes from one reaction channel to the next, which means that in the quantum theory there are no unique coordinates of the form in Equation 6 in which reactions are described. As a result, the simplicity with which one solves the classical n-body collision problem is lost in the quantum theory. In fact, the problem is so serious in the quantum theory that even some of the simplest reaction problems, i.e., atom-diatom collisions, are not readily solved in general.

The functions φ are not orthogonal in coordinate space, as it is shown in the Appendix A for the case of elastic collisions. Because of this, direct inversion of Equation 69 is not possible, but if certain conditions on φ_0 are fulfilled, as discussed in Appendix A, the coefficient A is

$$A = \frac{1}{(2\pi)^3} \int d^3 Q_{AB}\, e^{-ik_{AB}Q_{AB}}\, \varphi_0 \tag{72}$$

With the coefficient in Equation 68 the time evolution of the wave packet is completely determined. In particular, for $t \rightarrow \infty$, when all the dissociation processes have occurred, the wave packet is

$$\psi \sim \psi_{AB}(t) + \psi_{SC} \tag{73}$$

where ψ_{AB} is the undisturbed initial wave packet and ψ_{SC} is the scattered part, given by

$$\psi_{SC} = \sum_{CD} \frac{\psi_{CD}}{Q_{CD}} \int d^3k_{CD}\, e^{ik_{CD}Q_{CD} - iEt}\, f_{AB\rightarrow CD}\, A \tag{74}$$

From ψ_{SC} we obtain the wave packet scattering amplitude, which we define as

$$F_{AB\rightarrow CD} = \int d^3k_{AB}\, e^{iQ_{CD}(k_{CD} - k^0_{CD}) - i(E - E^0)t}\, f_{AB\rightarrow CD}\, A \tag{75}$$

where k^0_{CD} designates the mean momentum in the channel CD and E^0, the mean energy of the collision. The scattering cross-section is now defined as

$$\sigma_{AB\rightarrow CD} = \frac{k^0_{CD}}{k^0_{AB}} |F_{AB\rightarrow CD}|^2 \tag{76}$$

which is a simple generalization of the cross-section when φ_0 is the plane wave. It is essential to notice that the cross-section (Equation 76) is time dependent, which is a consequence of another quantity — the time delay, which was already defined in the classical theory. Here we will give its quantum definition, which is derived from the wave treatment rather than from the concept of trajectory.

Let the modulus of the initial wave packet φ_0 at $t = 0$ have a maximum at some initial separation Q^0_{AB} of the molecules A and B. From that moment we measure the time when the wave packet arrives, e.g., in the channel CD, at some point outside the scattering region. The coordinates of this point are the spherical angles θ and ϕ and some distance Q_{CD} from the center of the coordinate system. The time delay now is the difference between this time and the time which the same initial wave packet φ_0 would take to travel the distance Q^0_{AB} at the speed k^0_{AB}/μ_{AB} and the distance Q_{CD} at the speed k^0_{CD}/μ_{CD}, where μ_{AB} and μ_{CD} are the reduced masses of the molecules A and B, and C and D, respectively. It appears superficially that both the classical and the quantum definitions of the time delay are identical. This is not the case because there is a basic difference in the meaning of the two time delays. In the classical theory the time delay is well defined because it is associated with a single trajectory defined by its initial conditions. In other words, if we know in advance what the time delay for a given trajectory, describing a collision $A + B \rightarrow C + D$, is going to be, then that will be time that molecules C and D will be observed to arrive at the final points. In fact, this will be equal to the time delay plus the time of free flight of molecules A, B and C, D. No C and D molecules will be observed prior to or after this time. This conclusion also follows from the fact that a classical trajectory is uniquely determined by its initial conditions. However, in the quantum theory the position of the initial wave packet φ_0

determines only more or less (within a certain accuracy) the initial separation of the two molecules A and B, but not their impact parameter because it is assumed that φ_0 is independent of the coordinates perpendicular to the line connecting A and B. This indeterminacy means that there is an equal chance of finding molecules A and B for any other value of the impact parameter. Because of this A and B are scattered for various different values of the impact parameter, and since all of them are equally probable, we detect molecules C and D arriving at different times. Therefore, the initial wave packet φ_0, which had only a single maximum along the line of propagation, distorts after collision to a rather elongated wave packet along the radial coordinate. In fact, the scattered wave packet may not have just a single maximum, but it may have many maxima as shown in Figure 3b. Strictly speaking, the time delay has no meaning in quantum theory since we do not detect the arrival of a single pulse, but a sequence of pulses. Such pulses may be spread out in time so that it may not be possible to associate a single value of the time delay with φ_0 as there may be several values.

All the information about collision A + B → C + D comes from the wave packet scattering amplitude $F_{AB\to CD}(t)$ in which a central role is played by the scattering amplitude $f_{AB\to CD}$, which is obtained by solving the time-independent Schrödinger equation (Equation 76), with the boundary condition (Equation 71). In principle, therefore, the quantum scattering problem can be solved provided we know the solution of Equation 70, and we can calculate the integral (Equation 75). However, some of the beauty of the quantum theory lies in the theory of linear equations for which quite powerful analytical techniques for determining solutions to Equation 70 exist, which greatly helps the understanding of the essence of $F_{AB\to CD}$. In fact, quantum scattering theory is a description of the various ways of calculating $f_{AB\to CD}$ approximately and the derivation of its properties. In this sense theory is much richer than the classical theory of scattering which involves nonlinear equations about which relatively little is known. In what follows we will try to use some of the available techniques to analyze $F_{AB\to CD}$ and to find some of its properties.

The wave packet scattering amplitude is in the form of the integral representation (Equation 75) where A is given by Equation 72. If it is assumed that φ_0 is Equation 67 then A can be written in the more convenient form

$$A = \delta(k_x)\,\delta(k_y)\,\frac{e^{iQ_0(k_0 - k_z)}}{2\pi} \int dQ_z\, e^{i(Q_z - Q_0)(k_0 - k_z)} \omega_0(Q_z - Q_0) \tag{77}$$

where δ is the Dirac delta function. In Equation 77 we have omitted the index AB on k and Q. Therefore, the amplitude $F_{AB\to CD}$ is

$$F_{AB\to CD} = \int dk_z\, e^{iQ_{CD}(k_{CD} - k^0_{CD}) - i(E - E_0)t + iQ_0(k_0 - k_z)}\, f_{AB\to CD}\,\mathbf{A} \tag{78}$$

where

$$\mathbf{A} = \frac{1}{2\pi} \int d\,x\, e^{ix(k_0 - k_z)}\,\omega_0(x) \tag{79}$$

In general, the integral in Equation 78 has no analytical solution and has to be evaluated approximately. It is tempting to evaluate it by the stationary phase method, but for this it is necessary that the moduli of both $f_{AB\to CD}$ and **A** are not rapidly varying functions of k_z. Under certain conditions this is true for **A**. To find these conditions let us assume, for simplicity, that ω_0 is symmetric with respect to $x \to -x$ and that ω_0 rapidly goes to zero beyond $x = \Delta$, where Δ is called the width of the wave packet. The restriction that ω_0 is symmetric is not a serious obstacle since no additional insight is gained under more general assumptions.

We distinguish two cases; when the product $|k_0 - k_z| \Delta$ is very small and when it is very large. In the former case the amplitude **A** is

$$\mathbf{A} = \frac{1}{\pi}\int_0^\infty dx \cos x(k_0 - k_z)\, \omega_0(x) \sim \frac{1}{\pi}\int_0^\infty dx\, \omega_0(x) \tag{80}$$

which is large and nearly a constant. On the other hand, in the latter case the amplitude **A** is

$$\mathbf{A} \sim \frac{1}{\pi}\int_0^\infty dx \cos x(k_0 - k_z)\, \varphi(0) \sim 0 \tag{81}$$

and therefore, **A** is real and has a similar shape to ω_0, except that its width is inversely proportional to Δ, i.e., for $|k_0 - k_z| \ll \frac{1}{\Delta}$ the amplitude **A** is large while for $|k_0 - k_z| \gg \frac{1}{\Delta}$ it is negligible. **A** has a maximum at $k_z = k_0$.

If for the moment we neglect the scattering amplitude $f_{AB\rightarrow CD}$ and formally apply the stationary phase method to the integral in Equation 78 we find that the largest contribution to $F_{AB\rightarrow CD}$ comes from the region of k_z for which

$$\frac{d}{dk_z}(Q_{CD}\, k_{CD} - Et - Q_0 k_z) = \phi'(k_z) = 0 \tag{82}$$

When we take into account the fact that

$$k_{CD} = \left[2\,\mu_{CD}\left(\frac{k_z^2}{2\,\mu_{AB}} + E_A + E_B - E_C - E_D\right)\right]^{1/2} \tag{83}$$

then from Equation 82 we see that

$$t = \frac{Q_{CD}\,\mu_{CD}}{k_{CD}} - \frac{Q_0\mu_{AB}}{k_z} \tag{84}$$

The approximate value of the integral (Equation 78) is

$$F_{AB\rightarrow CD} \sim \left(\frac{2\pi}{\phi''}\right)^{1/2} A(k_z)\, e^{i\phi} \tag{85}$$

where ϕ is the phase of the integrand. The maximum of $F_{AB\rightarrow CD}$ coincides with the maximum of **A**, at $k_z = k_0$. Therefore, if the stationary point of the phase of the integrand is calculated at $k_z = k_0$ we obtain the time when the wave packet maximum reaches the observation point. This time is given by Equation 84 with k_z replaced by k_0. By noting that $Q_0 < 0$ we recognize that t given by Equation 84 is the time which the free wave packet requires to travel the distance Q_0 plus Q_{CD}. From our definition of the time delay its value in this case is zero. Therefore, with $f_{AB\rightarrow CD} = 1$ in Equation 78 the wave packet scattering amplitude represents a noninteracting wave packet.

The most important contribution to the integral (Equation 78) comes from the interval of k_z for which the product $(k_z - k_0)^2\,|\phi''|$ is small. This means that when

$$|k_z - k_0|\sqrt{|\phi''|} \gg 1 \tag{86}$$

the contribution of the integral to $F_{AB\to CD}$ is negligible. Within this region of k_z around $k = k_0$ the amplitude **A** should not change appreciably, otherwise the stationary phase method cannot be applied. It has been shown that the amplitude **A** does not change much if

$$|k_z - k_0| \ll \frac{1}{\Delta} \tag{87}$$

and, therefore the approximate value of $F_{AB\to CD}$, given by Equation 85, is accurate if

$$\Delta \ll \sqrt{|\phi''|} \tag{88}$$

In other words, the stationary phase method works only for a narrow wave packet and not in the limit of the incident plane wave, i.e., for $\varphi_0 = 1$. However, in the latter case

$$\mathbf{A} \sim \delta(k_z - k_0) \tag{89}$$

and the integral (Equation 78) must be evaluated according to the rules for the delta functions. Consequently,

$$F_{AB\to CD} = f_{AB\to CD} \tag{90}$$

which is a well-known result in time-independent scattering theory. In these circumstances it is not possible to define the time delay because the incident wave is infinitely long and has no obvious maximum. Mathematically, this manifests itself in the time-independent scattering amplitude (Equation 90).

In the discussion so far we have set aside any reference to $f_{AB\to CD}$, except in Equation 90. Formally, the integral (Equation 78) can be evaluated, with this amplitude included, if it is assumed that the modulus of $f_{AB\to CD}$ is a slowly varying function of k_z, in the vicinity of k_0. In such a case the time of arrival of the maximum of the wave packet is obtained from Equation 82, where the phase $\phi_{AB\to CD}$ of the amplitude $f_{AB\to CD}$ is also included. This time is

$$t = \frac{Q_{CD}\,\mu_{CD}}{k_{CD}} - \frac{Q_0\mu_{AB}}{k_z} + \frac{\mu_{AB}}{k_z}\frac{d}{dk_z}\phi_{AB\to CD} \tag{91}$$

and if the time of arrival of the free wave packet (Equation 84) is subtracted, we obtain the time delay[9,10]

$$\tau = \frac{\mu_{AB}}{k_z}\frac{d}{dk_z}\phi_{AB\to CD} \tag{92}$$

This formal approach to the derivation of the time delay and the wave packet scattering amplitude is legitimate under the condition (Equation 88) if $\phi_{AB\to CD}$ is also included in ϕ.

However, the time delay as defined in Equation 92, has some serious limitations not related to the use of the stationary phase method. In order to see this limitation, let us, for simplicity, assume that the scattering amplitude $f_{AB\to CD}$ is given by the sum of the two terms

$$f_{AB\to CD} = e^{i\phi} + e^{i\psi} \tag{93}$$

In such a case $F_{AB\to CD}$ represents the propagation of *two* wave packets with time delays proportional to $d\phi/dk_z$ and $d\psi/dk_z$, respectively. The apparent splitting of the incident wave packet into two (or several) is quite plausible; one need only recall that the incident plane wave represents collisions at all impact parameters, and it is just possible that for two given impact parameters the final states may be the identical. Therefore, the parts of the incident wave packet which scatter at these two impact parameters will be "deflected" into these two final states. Since there is no reason why the time delays of the two wave packets should be the same, they come out of the interaction region at different times. Examples of such a process will be discussed in this book. Here we mention only one; direct reflection and orbiting in elastic atom-atom collisions which have the same scattering angle but different impact parameters.

In the derivation of the expression for the time delay in Equation 92, however, we have assumed that the scattering amplitude is a product of its modulus and the phase factor, a standard form for complex numbers. The amplitude in Equation 93 thus can be represented as

$$f_{AB\rightarrow CD} = 2\cos\frac{(\phi - \psi)}{2}\, e^{i\frac{1}{2}(\phi+\psi)} \tag{94}$$

Such a time delay does not reflect reality since it associates only *one* time delay

$$\tau \sim \frac{1}{2}\left(\frac{\partial \phi}{\partial k_z} + \frac{\partial \psi}{\partial k_z}\right) \tag{95}$$

with a given transition. Furthermore, for the time delay in Equation 95, the modulus in Equation 94 is not a slowly varying function of k_z in the vicinity of k_0, and therefore the procedure used to derive Equation 95 is not a legitimate conclusion which also applies to Equation 92. Such uncritical use of Equation 92 may have, in some circumstances, values for the time delay which are not physically meaningful.

As we have noticed, the main cause of the inadequacy in the definition (Equation 92) is the fact that the scattering amplitude $F_{AB\rightarrow CD}$ cannot be represented in the form

$$f_{AB\rightarrow CD} = |f_{AB\rightarrow CD}|\, e^{i\,\arg(f_{AB\rightarrow CD})} \tag{96}$$

and at the same time ensure that its modulus is a slowly varying function of k_z near k_0. There are cases where this is true, but they are exceptions rather than rules. A more realistic parametrization is

$$f_{AB\rightarrow CD} = \sum_n f^{(n)}_{AB\rightarrow CD} \tag{97}$$

where we will call $f^{(n)}_{AB\rightarrow CD}$ partial scattering amplitudes. These are chosen in such a way that the modulus *is* a slowly varying function of k_z near k_0 for each one of them. There is no obvious way to realize such a parametrization, but one possibility is to use a semiclassical theory based on the Feynman and Hibbs[7] path integral techniques. The index n then runs over all classical trajectories which lead from the initial to the final states. However, such an approach is a hybrid of the classical and quantum theories, which has some drawbacks. In purely quantum theory the amplitudes with these properties are difficult to choose. The obvious choice is the partial wave method in which the index n corresponds to the total angular momentum of the system. Since this method is the standard technique in the quantum theory, we will briefly describe its essentials and limitations. In order to simplify the discussion we will consider the elastic scattering of two atoms. The results obtained will, however, also be valid in the general case.

For the elastic scattering of two atoms the solution of the Schrödinger equation should be asymptotically

$$\varphi \sim e^{ikz} + \frac{e^{ikr}}{r} f(\theta, \phi) \tag{98}$$

where r is the interatom distance. Obtaining a solution with such a boundary condition is a difficult task which can be simplified considerably if φ is represented as the series

$$\varphi = \frac{1}{r} \sum_{l,m} \varphi_{l,m}(r)\, \gamma_l^m(\theta, \phi) \tag{99}$$

where $\gamma_l^m(\theta, \phi)$ are spherical harmonics and $\phi_{l,m}$ are partial wave functions. It should be pointed out, and this will help in later discussions, that such an expansion is primarily introduced to help in the solution of the boundary value problem. This expansion allows physically meaningful justification, as the quantum number l can be associated with the angular momentum. There is absolutely no reason why this expansion is unique as this partial wave expansion (Equation 99) provides only one of many ways of obtaining the scattering amplitude $f(\theta, \phi)$. The way in which $f(\theta, \phi)$ is obtained is irrelevant, as long as one finds a convenient way of doing it. In nuclear physics the expansion (Equation 99) is convenient because only a small number of partial waves are required, but in the atom and molecule collisions problem hundreds of partial waves are needed, and therefore such an approach becomes unsatisfactory. It is retained because of its physical interpretation, but very often because of this it can obscure the real physical process. Inadequacies of the expansion (Equation 99) are more pronounced in inelastic and reactive collisions, but one should say that very few alternative approaches exist.

However, let us go back to the expansion (Equation 99). It is a straightforward matter to show that for a spherically symmetric atom-atom potential, the scattering amplitude is

$$f(\theta, \phi) = \frac{1}{2ik} \sum_{l=0}^{n} (2l + 1)\, P_l(\cos\theta)\, (S_l - 1) \tag{100}$$

where $P_l(z)$ is the Legendre polynomial and S_l is the S-matrix. The scattering amplitude is obtained from the radial solution $\varphi_l(r)$, which satisfies the equation

$$\varphi_l'' = \left[U(r) + \frac{l(l+1)}{r^2} - k^2\right] \varphi_l \tag{101}$$

together with the boundary conditions

$$\lim_{r' \to R} \varphi_l(r) = 0$$

$$\lim_{r \to \infty} \varphi_l(r) \sim e^{-ikr} - (-1)^l S_l e^{ikr} \tag{102}$$

The S-matrix S_l is unitary, i.e., it can be represented as

$$S_l = e^{i\delta_l} \tag{103}$$

where δ_l, the phase shift, is real. Therefore, the wave packet scattering amplitude (Equation 78) is

$$F = \frac{1}{2i} \sum_l (2l + 1) P_l(\cos\theta) \int \frac{dk}{k} e^{ir(k-k_0) - i(E-E_0)t + iz_0(k_0-k)} (S_l - 1) \mathbf{A} \qquad (104)$$

In this expression there are two integrals which arise from the factor $S_l - 1$ in the integrand. Considering the integral which involves S_l, then from stationary phase theory we find that the corresponding time delay for $k = k_0$ is

$$\tau_l = \frac{\mu}{k_0} \frac{d}{dk} \delta_l \qquad (105)$$

while for the other integral, (arising from the factor -1 rather than S_l) the time delay is zero. Hence, if t in Equation 104 is restricted to the value which corresponds to the time delay in Equation 105 for one of the partial waves, then for any other partial wave in Equation 104 the stationary point of the first integral is obtained from the equation

$$r - z_0 - \frac{k}{\mu} t_{l_0} + \delta_l' = 0 \qquad (106)$$

and in general its solution is different from $k = k_0$. The stationary point of the second integral is

$$k = \frac{r - z_0}{t_{l_0}} \mu \qquad (107)$$

and is independent of l. If we designate the value t_0 by

$$t_0 = \frac{r - z_0}{k_0} \mu \qquad (108)$$

which is the time for a free wave packet to travel between the initial and final points, then Equation 106 becomes

$$\tau_l - \tau_{l_0} = t_0 \left(1 - \frac{k_0}{k}\right) \qquad (109)$$

and Equation 107,

$$k = \frac{k_0}{1 + \frac{\tau_{l_0}}{t_0}} \qquad (110)$$

In the usual circumstances the value of t_0 is much larger than τ_{l_0}, and, therefore, Equations 110 and 107 give the same stationary point $k = k_0$. This means that the scattering amplitude F in Equation 104 is approximately

$$F \sim \frac{1}{2ik_0} \left(\frac{2\pi\mu}{t_0}\right)^{1/2} \mathbf{A}(k_0) \sum_l (2l + 1) P_l(\cos\theta) (S_l - 1) \qquad (111)$$

which is apart from a factor, the scattering amplitude in the stationary wave limit. As a consequence, we notice that if the source of the molecular beam and detector is far away from the interaction region there is no way to measure the time delay of the initial wave packet. We notice the arrival of only one very broad packet, because in the whole range of time delays, between zero and some maximum value, its intensity is the same, given by Equation 111. This is entirely due to the broadening effect of the free wave packets.

The previous analysis was approximate because it assumed from Equations 106 and 110 that the approximation $k = k_0$ follows. Strictly speaking, k in Equation 110, differs from k_0 by the amount

$$\Delta k = \frac{\tau_l - \tau_{l_0}}{t_0 + \tau_{l_0} - \tau_l} k_0 \tag{112}$$

which means that the integral in Equation 104, for the lth partial wave is proportional to

$$\int_{-\infty}^{\infty} dk \cdots \sim \mathbf{A}(k_0 + \Delta k) \tag{113}$$

Therefore, if $\mathbf{A}(k_0 + \Delta k)$ is small then the contribution of this integral to F is negligible and Equation 104 becomes

$$F \sim \frac{1}{2i} \sum_{l=l_0-l'}^{l=l_0+l'} (2l + 1) P_l(\cos\theta) \int_{-\infty}^{\infty} \frac{dk}{|k|} e^{ir(k-k_0) - i(E-E_0)t + iz_0(k_0-k) + i\delta_l(k)} \cdot \mathbf{A}(k) \tag{114}$$

where l' is the value of l for which $\mathbf{A}(k_0 + \Delta k)$ is small.

However, it has been shown earlier that the width of $\mathbf{A}(k_0 + \Delta k)$ is of the order $\Delta k \sim 1/\Delta$, and, therefore, the value of l' can be determined by using Equation 112, which gives

$$\frac{1}{\Delta} \sim \frac{|\tau_{l_0} - \tau_l'|}{t_0} k_0 \tag{115}$$

where $\tau_{l_0} - \tau_{l'}$ has been neglected compared to t_0. This means that the accuracy with which the time delay can be measured depends on several factors, but the crucial one is t_0, the travel time of the free wave packet. If we take into account Equation 108 then Equation 115 takes another form

$$|\tau_{l_0} - \tau_l'| \cdot \Delta \sim \frac{r - z_0}{E_0} \tag{116}$$

from where we also notice that higher accuracy is obtained for τ_{l_0} at higher collision energy as well as for large Δ. The last possibility is academic because in this case the time delay loses its meaning, as discussed earlier.

In the circumstances that l' in Equation 114 is finite, the wave packet scattering amplitude is still the sum over the partial waves which can, in fact, involve a large number of terms. This is because l can have a large absolute value which is small compared to l_0, as is generally the case for atom and molecule collisions. Therefore, the parametrization of the scattering amplitude $f(\theta, \phi)$ (based on partial wave decomposition) formally satisfies the conditions in the decomposition in Equation 97 but it does, however, have a drawback as it is fragmented.

In one extreme the scattering amplitude is not parametrized and passes over to another extreme where, in the partial wave decomposition, each term in the sum corresponds to part of a wave packet with a certain time delay rather than a single wave packet. Therefore, what is needed is a parametrization of $f(\theta, \phi)$ intermediate between the nonparametrized and the partial wave decomposition extremes.

II. RESONANCES AND LONG-LIVED STATES

All the properties of the long-lived states are, in principle, contained in the scattering amplitude $f_{AB \rightarrow CD}$. We say in principle, because not all of them have yet been found. However, there is one property which all the long-lived states possess; when such states are formed the phase of $f_{AB \rightarrow CD}$ changes very rapidly with the collision energy. This feature of the scattering amplitude follows from the definition of time delay (Equation 92). The cause of such phase behavior differs for different types of long-lived states, and therefore the object of the theory is to find all of the causes. This will enable us to classify such states and determine how they affect the cross-sections.

In order to solve the problem properly, it must be treated in two separate steps. The first step is to find a convenient expansion of the scattering amplitude, of the form in Equation 97, and analyze each term in the expansion separately. The reason why the scattering amplitude itself cannot be analyzed has already been discussed, and the conclusion was that the definition of time delay may lead to meaningless results. Here we encounter the first difficulty. What is the best way of defining the partial scattering amplitudes? The obvious choice is the partial wave decomposition, but as we have already seen, such an expansion is not necessarily the most convenient one. In the course of this book we will give an alternative to this expansion, which is much better for the analysis of long-lived states than the previous one.

After this step each partial scattering amplitude is analyzed in order to find possible long-lived states, i.e., the intervals of collision energy where the phase of these amplitudes changes very rapidly. It is also essential to find the source of such behavior in order to distinguish between various types of long-lived states. For this purpose there are several powerful techniques at our disposal, i.e., the theory of analytic functions, perturbation theory, and semiclassical theory. Although quite powerful, it is sometimes very difficult to apply any of them, and therefore the features of various types of long-lived states must often be described using simple models. Throughout the book we will see how these techniques work in particular circumstances.

In this section we will discuss in detail a particular kind of state, called a resonance, which has its origin in the analytic structure of the scattering amplitude. Strictly speaking, resonances, as we shall see, are not true long-lived states, but because they are long-lived states in some broader sense we include their discussion in the book. Also they have very simple descriptions in terms of the properties of the scattering amplitude, and therefore we will devote the whole section to their study. Other long-lived states will only be mentioned here in general terms. A more detailed study of these states will be developed in subsequent chapters. However, for atom and molecule collisions, resonances are less important than other long-lived states because they are only observed in narrow energy intervals, a feature unique to them. More will be discussed later, but here their general description will be presented.

Let us assume that we have parametrized the scattering amplitude according to Equation 97, say in terms of partial waves. A rapid change in the phase of one partial scattering amplitude may be due to the presence of a pole (of this amplitude) in the complex k-plane. The pole itself has no physical meaning since it is a consequence of the analytic structure of the solution of the Schrödinger equation in the complex k-plane with the boundary condition Equation 71. However, if such a pole is close to the real k-axis, its existence will be noticed as a rapid change of the scattering amplitude when the collision energy is scanned. It is obvious that the extent of this change will depend on how close this pole is to the real axis.

It should be noted here that description of certain phenomena, in terms of the complex quantities, i.e., the poles of the scattering amplitude, is not only peculiar to the quantum theory. Complex poles also appear in the classical theory when one solves the problem of the forced harmonic oscillator with a damping term. In this problem the amplitude of the oscillator has a complex pole in the frequency of the external harmonic force.[11]

Although the single pole approximation is an oversimplification of a true atom and molecule collision problem, we will use it in order to learn how such poles affect the wave packet. In general, each partial scattering amplitude has many poles, and their effect on the scattered wave packet depends on how they are grouped together. It will, however, be possible to learn something about this more complex general case from a single pole approximation.

It can be shown that the k-poles of the scattering amplitude cannot be real and that they are in the lower half of the k-plane, i.e., if k_p is a pole then $\mathrm{Im}(k_p) < 0$.[12] There are also imaginary poles, but for our purpose they are not important since they represent bound states. Therefore, in the vicinity of k_p we can write the nth partial scattering amplitude as

$$f^{(n)}_{AB \to CD} \sim \frac{R^{(n)}_{AB \to CD}}{k - k_p} \tag{117}$$

where $R^{(n)}_{AB \to CD}$ is a function of k. The corresponding wave packet scattering amplitude is

$$F^{(n)}_{AB \to CD} = \int d\,k\, e^{iQ_{CD}(k_{CD} - k^0_{CD}) - i(E - E_0)t + iQ^0_z(k_0 - k)} \frac{A\, R^{(n)}_{AB \to CD}}{k - k_p} \tag{118}$$

The form of the scattered packet, when a pole is present in the scattering amplitude, is obtained by calculating the integral Equation 118. The structure of this integral is the same as that of the ordinary wave packet, except that a pole is present in the integrand in Equation 118. A consequence of this difference is that Equation 118 cannot be integrated by a straightforward stationary phase method because in the vicinity of a pole both the phase and the modulus of the integrand vary very rapidly. This is obvious when the pole term in Equation 117 is written in a polar form

$$\frac{1}{k - k_p} = \frac{e^{i\tan^{-1}\frac{\gamma}{k - k_r}}}{[(k - k_r)^2 + \gamma^2]^{1/2}} \tag{119}$$

where γ and k_r are defined as $k_p = k_r - i\gamma$; $\gamma > 0$. When $k \sim k_r$ and γ are small, the modulus of Equation 119 changes rapidly contrary to the assumptions necessary for the application of the stationary phase method.

Here we will show how this integral can be calculated for a particular form of the initial wave packet. Three typical stages are involved: the moment of arrival of the wave packet in the scattering region, stationary scattering, and the end of scattering. The Gaussian form of the initial wave packet, which is often used,[13] is not good for this purpose, although it is the simplest for calculation of Equation 118. This represents a simple pulse, and therefore such a description does not allow all three stages to be disentangled. This becomes particularly evident in the stationary scattering limit. The wave packet in such a limit spreads in all

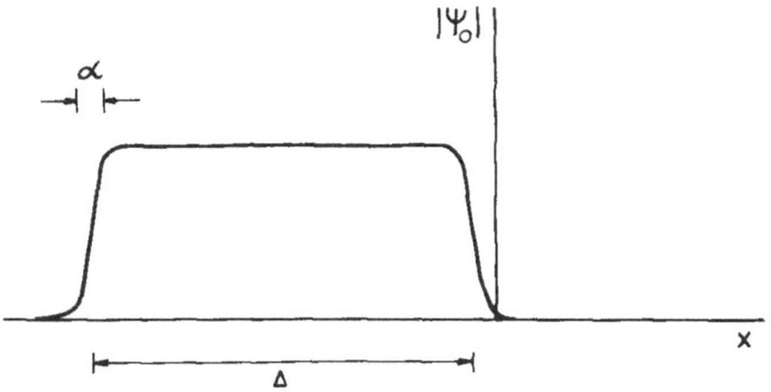

FIGURE 4. Wave packet for which three stages of scattering can be disentangled; moment of arrival, stationary scattering, and the end of scattering, α is width of the edges of the wave packet and Δ is its length

directions, and therefore the exact moment of scattering becomes undetermined. A much better choice is

$$\varphi_0(x) = \frac{e^{-\alpha x}}{1 + e^{-\alpha x}} - \frac{e^{-\alpha(x+\Delta)}}{1 + e^{-\alpha(x+\Delta)}} \tag{120}$$

which represents the wave packet shown in Figure 4. Δ is the length of the wave packet, and α is the width of its beginning and the end. By taking the limit $\Delta \gg \alpha$, it will be possible to analyze what happens to the scattered wave packet during all three scattering stages separately.

For this wave packet the amplitude **A** is

$$\mathbf{A} = \frac{1}{2i} \frac{1 - e^{-i\Delta(k_0 - k)}}{\mathrm{sh}\,\dfrac{k_0 - k}{\alpha}} \tag{121}$$

and the amplitude $F^{(n)}_{AB \to CD}$ is

$$F = \frac{1}{2i} \int d\,k\, e^{irk - i\frac{k^2 t}{2\mu} - iz_0 k} \frac{R(1 - e^{-i\Delta(k_0 - k)})}{\mathrm{sh}\,\dfrac{k_0 - k}{\alpha}} \frac{1}{k - k_p} \tag{122}$$

where all indexes and the nonessential phases have been omitted and it is assumed that $k_{CD} \sim k$, to simplify also the discussion. Setting $r = R_{CD}$ and $z_0 = Q_z^0$ also eases the notation.

There are two separate integrals in Equations 122; one is

$$F_1 = \frac{1}{2i} \int_{L_1} dk\, e^{irk - i\frac{k^2 t}{2\mu} - iz_0 k} \frac{R}{(k - k_p)\,\mathrm{sh}\,\dfrac{(k_0 - k)}{\alpha}} \tag{123}$$

and the other, which is designated by F_2, contains the factor $\exp[-i\Delta(k_0 - k)]$ in the integrand, and hence $F = F_1 - F_2$. The integration path L_1 in Equation 123 runs along the real k-axis, but must not contain the point $k = k_0$, otherwise the separation into F_1 and F_2

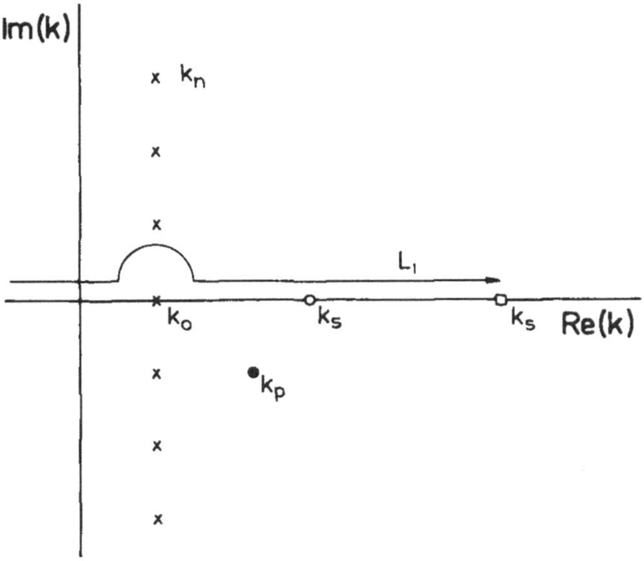

FIGURE 5. Characteristic points of the integrand in Equation 123. k_p is the resonance pole, k_s are stationary points of the phases of F_1 and F_2, and k_n are roots of the denominator of integrand.

is not possible. Therefore, L_1 is made to circumvent this point and, indeed, along this path $F_1 - F_2$ equals F which follows from the fact that $k = k_0$ is not a pole of the integrand of F.

The poles $k_n = k_0 + i\,\pi n$; $n = 0, \pm 1, \pm 2, \ldots$, designated by crosses, and the stationary point k_s of the phase of the integrand of F_1, designated by a circle, are shown in Figure 5. In the same figure we also show the integration path L_1.

For a given t the value of the stationary point k_s is obtained from

$$r - z_0 - \frac{k_s}{\mu} t + \frac{k_s}{\mu} \tau_s = 0 \tag{124}$$

where τ_s is time delay of the wave packet if f is replaced by R. Therefore, k_s must be calculated for different t values from Equation 124. In the simplest case it can be assumed that $\frac{d}{dk}\arg(R)$ does not change appreciably in the interval of k_s around k_0, and therefore

$$k_s = \frac{\mu}{t}(r - z_0) + \frac{k_0}{t} \tau_0 \tag{125}$$

where τ_0 corresponds to the time delay for $k = k_0$. More complicated behavior of τ_s in Equation 124 would not considerably alter our conclusions.

Poles of the integrand of F_2 coincide with those of F_1, but the stationary point of F_2 is shifted by $\Delta\frac{\mu}{t}$ with respect to k_s, the stationary point of F_1. In Figure 5, this stationary point is designated by a square. The position of the pole, k_p, with respect to k_0 is arbitrary, and in our discussion we will assume $k_r > k_0$. The case $k_r < k_0$ is treated in an analogous manner.

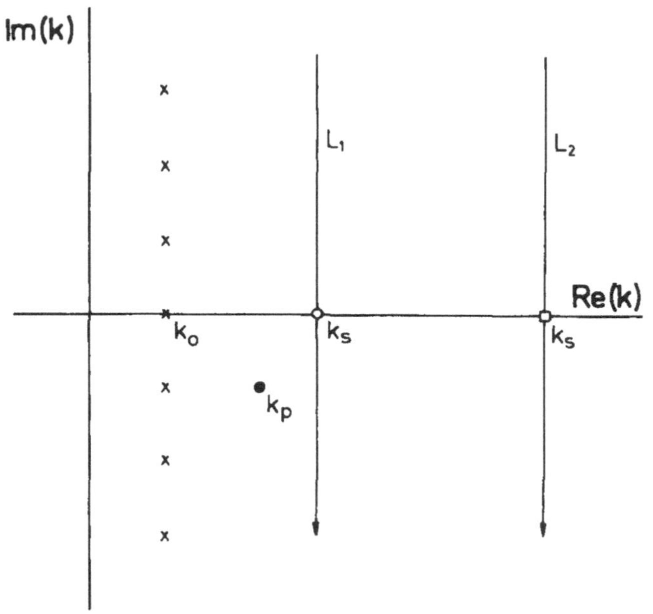

FIGURE 6. Transformation of integration path of Figure 5. L_2 is integration path for F_2.

The shape of the scattered wave packet F depends on the relative position of k_s with respect to k_p and k_0. If $k_s > k_r$ the integration path in Equation 123 can be distorted into the path shown in Figure 6. In such a case F_1 is

$$F_1 = -\pi\alpha \sum_{n=1}^{\infty} (-1)^n \frac{R}{k_n - k_p} e^{i(r-z_0)k_n - i\frac{t}{2\mu}k_n^2}$$
$$- \frac{R}{2} e^{i\frac{k_s^2 t}{2\mu} - \frac{k_0^2 \tau_0}{\mu}} \int_{-\infty}^{\infty} e^{i\frac{ty^2}{2\mu}} \frac{dy}{(k_s + k_p + iy)\,\mathrm{sh}\,\dfrac{(k_0 - k_s - iy)}{\alpha}} \qquad (126)$$

the first term in F_1 is the sum over the residues of the integral F_1 at $k = k_n$. Each term in the sum is of the order $e^{-\pi\alpha n\frac{t}{\mu}(k_s - k_0)}$, and since $k_s > k_0$ the sum can be neglected in F_1. When $k_s \geqq k_p$ the integral in Equation 126 is calculated by the stationary phase method, and thus F_1 becomes approximately

$$F_1 \sim \frac{1}{\sqrt{t}} \frac{1}{k_s - k_p} \frac{1}{\mathrm{sh}\left(\dfrac{k_0 - k_s}{\alpha}\right)}, \qquad (127)$$

which is small for all times when $(k_s - k_0) \geqq \alpha$ or for

$$t \ll \frac{1}{\alpha + k_0} [\mu(r - z_0) + k_0 \tau_0] \qquad (128)$$

In other words, immediately after $t = 0$ there is no scattered wave packet.

As the time increases k_s approaches k_r and the sum in Equation 126 increases, but stays small, more or less. The integral in F_1 also increases, but it cannot be estimated by the stationary phase method; it must be evaluated by another approximation. If $k_s \gtrsim k_r \sim k_0$ then $\mathrm{sh}\,\dfrac{k_0 - k_s - iy}{\alpha} \sim -\dfrac{\alpha + iy}{2\alpha}\, e^{(k_s - k_0)/\alpha}$ and F_1 is

$$F_1 \sim \frac{e^{-\frac{k_s - k_0}{\alpha}}}{\alpha} \int dy \frac{e^{i\frac{ty^2}{2\mu}}}{iy + k_s - k_p} \tag{129}$$

which has a solution in terms of the error function:

$$F_1 \sim \frac{\pi}{\alpha} e^{-\frac{k_r - k_0}{\alpha} - i\frac{t}{2\mu}(k_s - k_p)^2} \operatorname{erfc}\left[(k_s - k_p)\sqrt{\frac{t}{2\mu}}\, e^{-i\frac{\pi}{4}}\right] \tag{130}$$

This is an estimate of F_1, which is valid until t is such that $k_s = k_r$. The estimate of F_2 is the same as F_1 for $k_s \gg k_r$, since Δ is large. Therefore, F_2 is negligible and $F \sim F_1$. In conclusion we can say that up to the time when $k_s = k_r$ the intensity of the scattered wave packet steadily increases and its main contribution comes from the integral Equation 129. In the vicinity of $k_s \sim k_r$ the intensity increases very rapidly.

For the times when $k_0 < k_s < k_r$ the integration path L_1 of F_1 runs parallel to that in Figure 6, but goes between k_0 and k_r. The integral F_1 is similar to Equation 126 except that there is an additional term, which is its residue at the pole k_p so that

$$F_1 = -\pi \frac{R(k_p)}{\mathrm{sh}\,\dfrac{k_0 - k_p}{\alpha}} e^{i(r - z_0)k_p - it\frac{k_p^2}{2\mu}} + F_1^0 \tag{131}$$

where F_1^0 is Equation 126. This additional term represents a decaying amplitude in time, of the form $\exp(-t\gamma k_r/\mu)$. Together with F_1^0 it represents the time behavior of the scattered wave packet. Since three terms contribute to F_1 it is of interest to estimate their orders of magnitude. When this is done the amplitude F_1 is estimated as

$$F_1 \sim 0_p\left[\frac{\alpha}{k_0 - k_p} e^{-\gamma\frac{t}{\mu}(k_r - k_s)}\right] + 0_s\left(\frac{1}{\alpha}\right) + 0_n\left[e^{-\pi\alpha(k_s - k_0)\frac{t}{\mu}}\right] \tag{132}$$

where 0_p corresponds to the pole term; 0_s, to the integral; and 0_n, to the sum over k_n. The relative importance of the terms is quite complex, but in general it depends on time and the difference $k_0 - k_r$. If the latter is small then at least at those times when $k_s \sim k_r$ the pole term is dominant. As the time increases the significance of the pole term diminishes, and the sum in Equation 126 starts to become important. The integral in F_1^0 is small, on average, but may be larger than the sum when $k_s \sim k_r$. In conclusion, we can say that in this interval of time the amplitude of the scattered wave packet increases, but later its behavior becomes quite complex. However, since $k_0 \sim k_r$ the significance of this time interval is not great, and, as in the previous time interval, the value of the integral F_2 is negligible.

In the next stage of scattering, when $k_s < k_0$, the integration path L_1 runs parallel to that in Figure 6, but is now to the left of $k = k_0$. This time there is contribution of the pole k_0 in F_2 because L_2 must cross k_0 when shifted into its proper position. In the previous time periods the integration path L of F went over $k = k_0$ to reach stationary point k_s, and therefore the integration path L_2 could have been shifted to the right, without crossing the point $k = k_0$. However, when k_s is now to the left of k_0, the integration path cannot be shifted to the right without crossing $k = k_0$.

The integral F_1 is the same as Equation 131, but the sum in F_1^0 runs over negative n values. On the other hand, the value of the integral F_2 is not now nonnegligible since its residual at $k = k_0$ is

$$F_2 = -\pi\alpha \frac{R(k_0)}{k_0 - k_p} e^{irk_0 - \frac{ik_0^2}{2\mu}t - iz_0k_0} \tag{133}$$

The amplitude F now has four components; three from F_1 and one from F_2. Some of the components from F_1, i.e., the sum and the integral in F_1^0 are ngeligible. The leading two components give

$$F \sim \frac{\pi\alpha R}{k_0 - k_p} e^{i(r - z_0)k_0 - i\frac{k_0^2 t}{2\mu}} \left[1 - e^{i\frac{t}{2\mu}(k_p - k_0)(2k_s - k_p - k_0)}\right] \tag{134}$$

where the approximation $\mathrm{sh}\frac{(k_0 - k_p)}{\alpha} \sim \frac{k_0 - k_p}{\alpha}$ has been made.

In this time interval, F asymptotically approaches a stationary value exactly equal to that as the incoming wave packet is a plane wave, with the wave number k_0. As is observed from Equation 134, if the cross-section is measured long after k_s is equal to k_p, it will have a constant value (in time) and will be quite large if γ is small and $k_0 \sim k_r$. Away from this value of k_0 the amplitude, and thus also the cross-section, decreases to much smaller values. Therefore, if the cross-section is scanned around $k_0 = k_r$ at the time when F is stationary, the cross-section will have a very large, rapidly changing amplitude, and the width in this interval is γ. From Equation 134 the rate at which the amplitude reaches the stationary value is obtained. The modulus of the exponential term in the brackets of Equation 134 is $\exp\left(-\frac{\gamma t}{\mu}(k_r - k_s)\right]$ from which it follows that this rate is governed by γ; the smaller γ, the longer it takes to reach the stationary amplitude.

If Δ is sufficiently long, then after the stationary phase, the last stage of scattering is reached, the arrival of the end of the wave packet. In this stage the stationary point k_s is far away from k_0 to the left, while the stationary point of F_2 is near k_0 and also to the left. When the integration path L_1 and L_2 are changed so that they run parallel to those in Figure 6, the amplitude F_1 becomes negligible. The amplitude F contains three terms (arising from F_2) similar to those in Equation 131, except that r is formally replaced by $r + \Delta$. If F_1^0 in Equation 131 is neglected, then F is approximately

$$F \sim F_2 \sim \pi\, e^{i(r + \Delta - z_0)k_p - it\frac{k_p^2}{2\mu}} \frac{R}{\mathrm{sh}\,\frac{(k_0 - k_p)}{\alpha}} \tag{135}$$

which represents a decaying amplitude, and it is all that is left of the wave packet. If γ is small, then F can stay appreciably large long after the nonscattered wave packet has passed the observation point. The time when the end of such a wave packet reaches the observation point occurs when the stationary point of F_2 coincides with k_0.

The decay term of Equation 135 is $\exp\left(-\frac{tk_r\gamma}{\mu}\right)$, which takes a more familiar form if instead of the pole k_p one considers the pole in the energy variable. If we define $E_p = \frac{k^2{}_p}{2\mu} \equiv E_r - i\Gamma$ then $F \sim \exp(-t\Gamma)$, and we see that the rate of decay of F is equal to the inverse of the width of the rapid change of the cross-section in the stationary phase of scattering (when the collision energy is scanned).

Such behavior of the scattered wave packet, which is caused by a pole in the scattering amplitude, is analogous to the phenomenon called resonance. Resonances are found in many systems i.e., in electronic circuits or mechanical oscillators, and they respond to an external pulse in exactly the same way as in the scattering case described above. Because of this analogy we say that the behavior of the scattered wave packet is caused by a resonance, and this resonance is represented by a pole in the scattering amplitude. The imaginary part of this pole, γ, measures the efficiency with which the system achieves the steady-state resonance energy; if γ is small it takes a very long time to achieve the steady state, while if it is large, the time is short. Physically, if the collision momentum k_0 is equal to the resonance momentum, k_r, the amplitude of the scattered wave packet in a steady state is proportional to γ^{-1}, and if γ is small, this amplitude is very large.

This means that the probability of finding scattered particles is greatly enhanced at the resonance energy, and since the incoming probability flux is normalized to unity, it takes time to "pump" sufficient flux into the system for a significant scattered probability to be achieved. This has analogy with the forced harmonic oscillator, damped by friction proportional to velocity. The amplitude of the oscillator in the steady state (and the resonance frequency) is proportional to the inverse of the friction coefficient. If friction is small, this amplitude is large, and a large amplitude means that a lot of energy is stored in the oscillator. Since the external force can only deliver a fixed amount of energy per unit time, it will take a long time to pump this energy into the oscillator. In other words, the smaller the friction is, the longer it will take for the oscillator to achieve a steady-state situation.

In the collision problem this also means that the resonant steady state will not be observed if the length of the incident wave packet is shorter than the time to achieve the steady state multiplied by the velocity k_0/μ. However, in most scattering experiments the incident wave packet is sufficiently long so that even a very narrow resonance can achieve the steady state before the end of the wave packet arrives.

When the end of the wave packet has passed, the steady state decays because there is no incoming flux. The time dependence of this decay is the same as the time dependence to achieve the steady-state situation, except the end limits differ.

After the analysis of resonances we can ask a crucial question: is a resonance a long-lived state? The reason why one should call a resonance the long-lived state is that one observes the probability flux long after the (assumed) end of the wave packet has gone. On the other hand, one can also say that in the presence of a resonance the scattered wave packet has been distorted from the incident shape so much that its true end is where the decay curve is almost zero. From this point of view a resonance is not a long-lived state.

In order to resolve this dilemma, let us go back to what we have defined as a long-lived state. If the time taken by the wave packet to arrive at the observation point is longer than that expected for a free wave packet, the delay is attributed to the formation of a long-lived state between the incident particle and the target. In other words, the long-lived state causes a positive time delay. When a resonance is present we notice from Equation 124 that the time of arrival of the wave packet is given by Equation 125 when $k_s \equiv k_0$, and therefore the time delay is

$$\tau_0 = \frac{\mu}{k_0}\frac{d}{dk_0}\arg(R) \tag{136}$$

which does not contain parameters of the pole. This means that resonance does not cause a time delay, and therefore strictly speaking, it is not a long-lived state. In fact, the simplest way to demonstrate that a resonance does not cause a positive time delay is with a short wave packet, i.e., one for which Δ is small. In such cases resonances do not have time to form, and we only notice that the wave packet is delayed by Equation 136. However, if we take into account formation of a resonance, then obviously the time delay in Equation 136 is prolonged by an amount of the order Γ^{-1}, which is often called the resonance lifetime. In this respect one can talk of resonances as long-lived states in a broader sense.

One resonance model which we have discussed is not a very good approximation for atom and molecule collision. In such collisions, in fact, there may be a large number of resonance, and thus the single resonance approximation may often be misleading. in order to see what happens to the scattered wave packet when only a few poles are present in $f_{AB \to CD}$, it is sufficient to analyze two-pole approximation, since generalization to the more general case is straightforward. We write

$$f^{(n)}_{AB \to CD} = \frac{R^{(n)}(1)}{k - k_1} + \frac{R^{(n)}(2)}{k - k_2} \tag{137}$$

and for simplicity we will assume that both poles are to the right of k_0. If we assume, as before, that the derivative $\frac{d}{dk}$ arg $[R^{(n)}(m)]$; m = 1, 2 does not change appreciably in the interval around $k = k_0$, then we distinguish two typical cases: (1) the time delays for two amplitudes, R, in Equation 137 are almost the same of (2) they differ by a large amount, i.e., $\tau_1 \geqq \tau_2$.

The scattered wave packet is obtained from Equation 122 and because of the parametrization in Equation 137, it splits into two parts, which we can write as

$$F(m) = \frac{1}{2i} \int_{m=1,2} dk\, e^{ikr - ikz_0 - i\frac{k^2 t}{2\mu}} \frac{R(m)[1 - e^{-i\Delta(k-k_0)}]}{sh\left(\frac{(k_0 - k)}{\alpha}\right)} \frac{1}{k - k_m}; \tag{138}$$

so that F = F (1) + F (2). Each component of F can now be analyzed separately, in the same way as in the one-pole case. Therefore, the scattered wave packet is a superposition of two wave packets, where each has the shape discussed earlier for the single-pole case. The only difference lies in the arrival times of two wave packets. If $\tau_1 \sim \tau_2$ then they arrive simultaneously and they are very difficult to distinguish. On the other hand, if $\tau_1 \geqq \tau_2$ then the wave packet F(2) with the time delay τ_2 arrives first, long before the wave packet F(1). In principle, these two wave packets are observed separately. It is obvious that when more poles are present the analysis of the scattered wave packet becomes more complicated.

So far the analysis has been restricted to a single partial scattering amplitude. The choice of this amplitude was left for later discussion, and this is the moment to say a few words about it. The most obvious choice is the partial wave decomposition of the scattering amplitude in which case the index n corresponds to the total angular momentum of the system. In such a case, if a pole is present in one partial wave scattering amplitude, one also expects a pole to be present in the neighboring partial wave amplitudes, though for a slightly different complex momentum, k. This is a consequence of the theorem which says that the derivative, $\frac{dk_p}{dn}$, is finite. The proof of this theorem will be omitted since throughout this book it will be explicitly shown that this theorem is true. The entire scattered wave packet is, therefore, a sum of the terms in Equation 118, where now the pole also has an additional index which corresponds to a particular partial wave where it is found. The difficulties with such an expression for the wave packet are the same as those discussed in the previous section.

Another, more convenient choice of the expansion Equation 97 is based on *Complex Angular Momentum Theory*, which is much better suited to the analysis of long-lived states than is partial wave decomposition. The theory has been developed by Sommerfeld and Watson[14] for electromagnetic scattering processes, and later it was extended by Regge[15] and Alfaro and Regge[16] to quantum scattering processes, particularly for the case of nuclear and elementary particle scattering.[17] In its original version the theory cannot be applied to atom and molecule collision because of the implicit assumption that particles interact by a potential which does not have very repulsive core. By a very repulsive core, it is meant that the potential at the distances which are not close to the origin is very large compared with the collision energy. A typical collision energy is 1 eV for neutrals and 10 to 100 eV for ions, but the Coulombic repulsive potential between atoms at short distances may be of the order of keVs or even MeVs. Therefore, for all practical purposes it can be assumed that a typical atom/molecule potential has a hard core. For such a potential Regge's complex angular momentum theory cannot be applied in a straightforward way, but it can be modified in order to take into account the hard core. This theory will be discussed in detail later.

Chapter 3

SEMICLASSICAL THEORY OF SCATTERING

So far we have demonstrated how the classical and quantum theories treat atom/molecule collisions. Strictly speaking this is a quantum problem, but one in which we already notice signs of difficulties, such as a large number of partial waves necessary to represent the scattering amplitude. This and other difficulties, which we will review, are specific to atom/molecule collisions, and they are attributed to what we will call their semiclassical nature. Usually what one means by this is gauged in terms of Planck's constant h and is described as the limit of "small h". In fact, behind this is the observation that for quantum systems, as mass increases, systems start to show classical properties. Since the ratio h^2/m always appears, this is equivalent to taking the limit $h \rightarrow 0$. The most immediate manifestation of this semiclassical limit, in the quantum theory, is that the wave functions rapidly oscillate over the range of the potential. For example, an iodine atom with collision energy of 1 eV has a wave length of 0.1 Å, which is much shorter than the range of a typical potential which is usually a few angstroms. Therefore, what we look for is a theory which will be valid in the transition region where "h is small but not zero". We can designate the classical theory limit by $h = 0$.

There are basically two approaches to this limit. In one, one starts with "large h" and approaches the limit $h \rightarrow 0$, and in the second, one starts from $h = 0$ and approaches the small $h \sim 0$ limit. The first approach starts from the quantum theory and obtains the solution for the short wavelength approximation. This approximation does not require any further assumption, but finding the solution very often depends on ingenuity and mathematical skill. The main mathematical tool is often the theory of asymptotic solutions of equations and asymptotic expansions.

The second approach starts from classical theory, or the trajectory concept, and generates the quantum effects. Here, however, a few other postulates or assumptions are necessary because the classical theory in itself does not contain the concept of interference. These assumptions are carried over from the analogy with rays, perpendicular to a wave front, which obey the classical equations of motion. In a sense this approach can be referred to as the real semiclassical theory, because it combines the trajectory and interference concepts from classical and quantum theory, respectively.

We will first review some basics of the short wavelength approximation. The Schrödinger equation for an N particle problem is a multidimensional partial differential equation of the second order which can be written as

$$-\frac{\hbar^2}{2}\sum_{i=1}^{N}\frac{1}{m_i}\left(\frac{\partial^2 \psi}{\partial x_i^2}+\frac{\partial^2 \psi}{\partial y_i^2}+\frac{\partial^3 \psi}{\partial z_i^2}\right)+V\psi = E\psi \qquad (139)$$

where m_i is the mass of ith particle. Solving this equation for the scattering boundary conditions is, in general, a formidable task. The various solutions are generally based on some intuition about the form of the solution, but even then difficulties may appear which are almost insurmountable. Therefore, we will consider a problem which appears to be simple, deceptively so. We will assume that before and after collision only two particles (molecules) are free, i.e., the scattering can be written as $A + B \rightarrow C + D$.

At this point we should make a distinction between two kinds of collisions; those which involve only one intermolecular potential (i.e., electrons are not excited in the collision) and those which involve electron excitation. In the discussion of the short wavelength approximation this distinction must be drawn because there is a qualitative difference between these collision events. In the former case Equation 139 contains only masses which correspond to the nuclei. On the other hand, when electrons are excited then some of the masses in Equation 139 correspond to electrons. Note that the limit $h \rightarrow 0$ only refers to the nuclear masses and not those of the electrons. The reason for this is straightforward. If the electron masses change, then the intermolecular potential must also change. Therefore, it is quite unrealistic to expect that the limit $h \rightarrow 0$ has a physical meaning for electrons. On the other hand, the masses of the nuclei can be changed, in principle, without much problem; one only has to add enough neutrons to them. In any case, since in nature there are different isotopes, the "variation of h" has already been realized. We will see later what the consequences of applying these two kinds of limits are.

Although the process $A + B \rightarrow C + D$ is the simplest collision case when clusters of particles are involved, this can be solved only with great difficulties if the particles which constitute A and B rearrange. We have seen the reason for this in the quantum theory discussion. Nevertheless, sometimes such events can be modeled by processes where there is no rearrangement. For example, in the charge exchange process $A + B \rightarrow A^+ + B^-$ which is, strictly speaking, a rearrangement collision, it can be assumed that the nuclear motion takes place on two potential surfaces, involving only *one* set of coordinates. Because such a model is possible and because of the difficulties with processes involving rearrangement, we will simplify our assumptions even further by assuming that no rearrangement occurs. In such a case the collision $A + B \rightarrow C + D$ can be described by one set of coordinates, and note that one coordinate is unique. This is the separation between the center of mass of A and B and the center of mass of C and D. We will designate this coordinate by r. In general, then, the equation for the system can be written as

$$\left(-\frac{\hbar^2}{2\mu}\Delta + T_A + T_B + V\right)\psi = E\,\psi \tag{140}$$

where μ is the reduced mass of the systems, i.e., $\mu = m_A m_B/(m_A + m_B)$. The kinetic energy operators of A and B (also of C and D, since there is no rearrangement) are T_A and T_B, respectively, while V is the potential. If T_A and T_B also refer to electrons, then V is the Coulomb interaction of all particles, but if T_A and T_B refer only to nuclei, then V is a single Born-Oppenheimer potential. The potential V can be parametrized as

$$V = V_A + V_B + V_{AB} \tag{141}$$

where V_A and V_B are the asymptotic values of V when $r \rightarrow \infty$, i.e., V_A and V_B are the potentials of isolated A and B species. In such a case $V_{AB} \rightarrow 0$ for $r \rightarrow \infty$.

The Schrödinger equation (Equation 140) can be solved in a closed form for $r \rightarrow \infty$. It has a solution

$$\psi = \psi_{n_A}\,\psi_{n_B}\,\varphi; \quad r \rightarrow \infty \tag{141a}$$

where ψ_{nA} and ψ_{nB} are the solutions of

$$(T_A + V_A)\,\psi_{n_A}(r_A) = E_{n_A}\,\psi_{n_A}(r_A)$$

$$(T_B + V_B)\,\psi_{n_B}(r_B) = E_{n_B}\,\psi_{n_B}(r_B) \tag{142}$$

Since the ψs represent bound states, $E_A, E_B < 0$, and the indexes n_A and n_B refer to sets of quantum numbers which describe the two states. The wave function φ is then the solution of

$$-\frac{\hbar^2}{2\,\mu}\Delta\,\varphi = (E - E_{n_A} - E_{n_B})\,\varphi \tag{143}$$

and depending on the sign of $E - E_{n_A} - E_{n_B}$ it is either the plane wave or increasing function of the coordinate. In the first case we talk about open channels, while in the second case we talk about closed channels.

One plausible way of solving Equation 140 is to assume that both ψ_{n_A} and ψ_{n_B} form a complete set of functions and expand ψ in such a set. It is only an assumption that ψ_{n_A} and ψ_{n_B} constitute such a complete set, but it is a useful one. However, even if this assumption is indeed true, there may be cases where such an expansion is not of much use. For example, if there are only a few eigenstates of Equation 142 the expansion of ψ may require the use of unbound solutions, which involve difficulties. In other circumstances, as shown shortly, the expansion in ψ_{n_A} and ψ_{n_B} may be poorly convergent.

Nevertheless, let us assume that we have made the expansion

$$\psi = \sum_{n_A, n_B} \psi_{n_A} \psi_{n_B} \varphi_{n_A, n_B} \tag{143a}$$

where the indexes run over both the open and closed channels. The expansion in Equation 140 is now replaced, and after projection on the channel (m_A, m_B) we obtain

$$-\frac{\hbar^2}{2\mu} \Delta \varphi_{m_A, m_B} + \sum_{n_A, n_b} \langle m_A, m_B | V_{AB} | n_A, n_B \rangle \varphi_{n_A, n_B} = (E - E_{n_A} - E_{n_B}) \varphi_{m_A, m_B} \tag{144}$$

where

$$\langle m_A, m_B | V_{AB} | n_A, n_B \rangle = \int dV_A \, dV_B \, \psi_{m_A} \psi_{m_B} V_{AB} \psi_{n_A} \psi_{n_B} \tag{145}$$

This is a set of coupled differential equations in one coordinate **r**, called the multichannel equations. They very much resemble the single particle equation except that there are many of them, and all mutually coupled.

This set can be written in a more compact form

$$\Delta \varphi = U \varphi - K^2 \varphi \tag{146}$$

where φ is a column vector, U is the matrix of the elements in Equation 145, and K^2 is a diagonal matrix

$$k^2_{n_A, n_B} = k^2 - k^2_{n_A} - k^2_{n_B} = \frac{2\mu}{\hbar^2} (E - E_{n_A} - E_{n_B}) \tag{147}$$

In the limit $\hbar \rightarrow 0$ we notice that both U and K^2 are of the order $0(\hbar^{-2})$ and, therefore, they diverge. Special attention should also be paid to $k^2_{n_A}$ and $k^2_{n_B}$. If ψ_{n_A} corresponds only to the electronic state, then E_{n_A} does not depend on the Planck constant, because it was assumed that the masses of electrons did not vary. In this case $k^2_{n_A}$ also depends on $\hbar$ as $0(\hbar^{-2})$. On the other hand, if ψ_{n_A} is a nuclear wave function, then E_{n_n} may involve both vibrational and rotational energy. Both are quantized and are of the orders

$$\begin{aligned} E_{vib} &\sim \hbar\omega(n + 1/2) \sim 0(\hbar) \\ E_{rot} &\sim \hbar^2 j^2 \sim 0(\hbar^2) \end{aligned} \tag{148}$$

Thus, $k^2_{n_A}$ is of the order $0(h^{-1})$ or $0(h^0)$ depending on whether the molecule A is vibrating or not. Therefore, in the limit $h \rightarrow 0$ the number of open channels does not increase for the

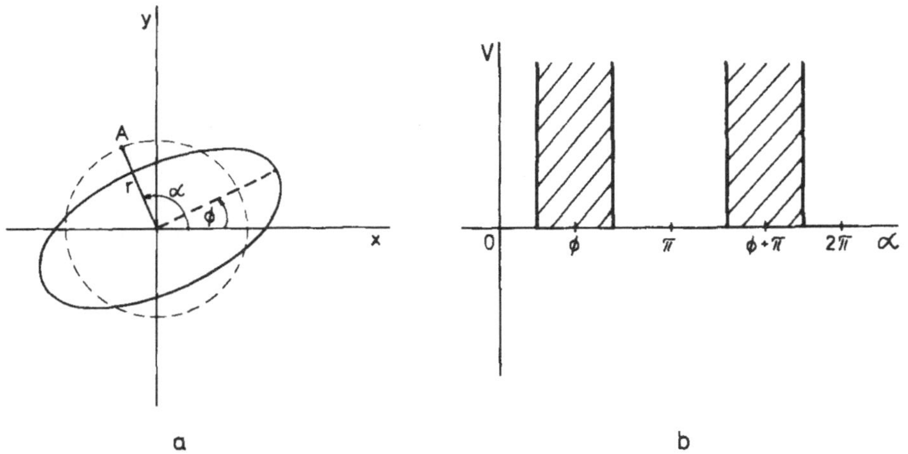

FIGURE 7. (a) Contour of a linear molecule as the incoming atom A "sees" it. (b) Atom-molecule potential along the broken line circle in (a). The details of this potential are neglected.

electronic energy though it increases as h^{-1} for the vibrational energy and as h^{-2} for the rotational energy.

In the simple case of a single channel Equation 144 can be solved in the limit $h \to 0$, and it is known as the WKB solution.[6] In Appendix B we show how this can be done. However, a similar approach for more than one channel is very difficult.[18] Various other ways should be investigated to find such a solution, but the question is if it is at all possible to find it. Besides the problem of finding such a solution, there is another drawback of the set of Equations 144 which is quite serious. This drawback is inherent in atom and molecule collisions, and it is essential to consider it.

Atom and molecule potentials are strongly repulsive at short internuclear distances, so much so that below certain distances they can be considered to have a hard core. Therefore, we can talk about the shape of a molecule as the other molecule "sees" it. For example, if an atom is approaching a molecule, which contains a linear row of atoms, it will "see" it as a cigar-shaped object. In the plane defined by the axis of the molecule and the position of the atom, this object may appear as shown in Figure 7a, where A is position of atom.

The potential V_{AB}, which enters Equation 145, is a function of the relative orientation of A and B, and therefore when the molecule is rotated by 2π (in Figure 7a this path is indicated by a circle) for fixed distance r and fixed atomic position, it will appear in Figure 7b. Of course, when r is greater than the extension of the "hard core" there will be no such infinite walls around $\alpha = \phi$ and $\alpha = \phi + \pi$. Most of transition, however, occurs when the atom hits the hard core of the molecule; therefore, the short distance region of potential V_{AB} is in fact the most important for collision. However, the matrix elements of Equation 145 are the average of V_{AB} and include integration of the potential as shown in Figure 7b. Therefore, if we attempt to write the solution of Equation 146 as an expansion

$$\varphi = \frac{1}{r} \sum_{l,m} Y_l^m(\theta, \phi)\ \omega_{l,m}(r) \tag{149}$$

then the equations for $\omega_{l,m}$, which are also the multichannel equations of the type in Equation 146, will have coupling elements which are very large. Most of the contribution in these elements comes from the repulsive region in Figure 7b, and this means that the set (Equation 146) is nonphysical because it has coupling elements which may have arbitrary values, since the value of V_{AB} around $\alpha = \phi$ can have any value. This region of the potential cannot be reached by the atom anyway. The nonphysical character of Equation 146 comes from our choice of expansion ψ into the asymptotic states. This expansion also includes the asymptotic angular functions, which are in fact the main cause of the problem. When the atom is far away from the molecule all the space around it is essentially accessible, and therefore these functions have the form of spherical harmonics. However, as the atom closes in on the molecule, some parts of space are no longer accessible, and therefore the functions should become zero in these regions.

Let us assume a different type of expansion than that in the asymptotic states. Let us solve the problem at each separation $\mathbf{r}$

$$(T_A + T_B + V)\,\psi_n(r_A, r_B; \mathbf{r}) = E_n(\mathbf{r})\,\psi_n(r_A, r_B; \mathbf{r}) \tag{150}$$

where E_n now depends parametrically on r. Then we can write for ψ

$$\psi = \sum_n \psi_n(r_A, r_B; \mathbf{r})\,\varphi_n(\mathbf{r}) \tag{151}$$

where, as before, n designates a set of all indexes which determine the eigenvalue E_n (r). The expansion of the type in Equation 151 is also known as the Born-Oppenheimer or adiabatic expansion. When electrons are involved it is much more suitable than the previous expansion. Expansion in the asymptotic states means that as two atoms approach each other, their electronic structures do not change much from their separated forms. It is obvious that this is not a realistic picture, and therefore for electrons the adiabatic expansion is used.

If the expansion Equation 151 is replaced in Equation 140 then we obtain a set of equations for $\phi_m(\mathbf{r})$

$$\Delta\varphi_m + \sum_n \left(2\langle\varphi_m|\nabla|\varphi_n\rangle\,\nabla\varphi_n + \langle\psi_m|\Delta|\psi_n\rangle\,\varphi_n\right) = [k_m^2(\mathbf{r}) - k^2]\,\varphi_m \tag{152}$$

The multichannel equations are now more complicated, and in order to justify their use we have to estimate the matrix elements in Equation 152. It can be shown that

$$\langle\psi_m|\nabla|\psi_n\rangle = \frac{\langle\psi_m|\nabla V|\psi_n\rangle}{E_n - E_m} \tag{153}$$

and

$$\langle\psi_m|\Delta|\psi_n\rangle = \nabla\langle\psi_m|\nabla|\psi_n\rangle + \sum_l \langle\psi_l|\nabla|\psi_m\rangle\,\langle\psi_l|\nabla|\psi_n\rangle \tag{154}$$

which means that Equations 153 and 154 are related.

The estimate of Equations 153 and 154 depends essentially on E_n, therefore we should have at least some qualitative idea of its behavior. In general, $E_n(\mathbf{r})$ depends on the orientation of the vector $\mathbf{r}$, i.e., E_n is not a spherically symmetric function. At long distances it represents the binding energies of A and B, but at short distances it is strongly repulsive, i.e., it represents the hard core. If A and B are only electronic degrees of freedom, then E_n is independent of $\hbar$. If A and B are molecules with only the nuclear coordinates, then E_n as explained earlier, depends on $\hbar$. In the former case $<\psi_m|\nabla|\psi_n>$ and $<\psi_m|\Delta|\psi_n>$ do not depend on $\hbar$, but in the latter case they do, but as an inverse power, i.e., $\hbar^{-1}$. As the result of this, when only the electrons are involved, we can obtain the approximate solution of Equation 152 from

$$\Delta\,\varphi_m^0 = [k_m^2(r) - k^2]\,\varphi_m^0 \tag{155}$$

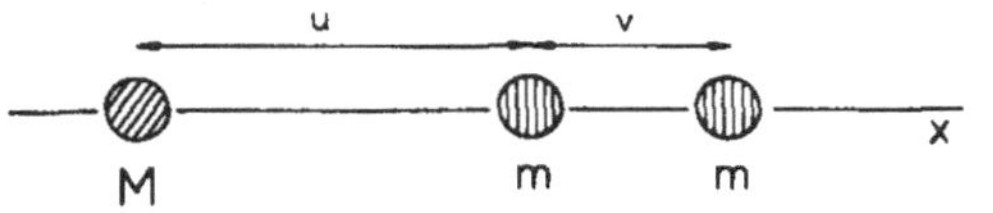

FIGURE 8. Coordinates for collinear collision of particle M with two particles m which are bound in infinite square well.

because the term ϕ_m in Equation 152 is of the order $\hbar^{-1}$ while the other term containing $<\psi_m|\Delta|\psi_n>$ is of the order $\hbar^0$. Both these terms are small compared to the right-hand side of Equation 152, which is of the order $\hbar^{-2}$

In the simplest case when $k_m^2(r)$ is spherically symmetric, i.e., when two atoms are colliding, ϕ_n is

$$\varphi_n = \frac{1}{r} \sum_{l,\mu} Y_l^{\mu}(\theta, \phi)\, R_{n,l,\mu}(r) \tag{156}$$

and the solution of Equation 152 can be written in the form of the Volttera integral equation for $R_{n,l}$

$$R_m = R_m^0 + \eta \sum_n \int_{r_0}^{r} K_m(r, r')\Big(2\langle\psi_m\Big|\frac{\partial}{\partial r}\Big|\psi_n\rangle R_n' + \langle\psi_m\Big|\frac{\partial^2}{\partial r^2}\Big|\psi_n\rangle\Big) \tag{157}$$

where we have omitted reference to the lth partial wave. The function $K_m(r, r')$ is the kernel, and η is a coefficient which is independent of r and is of the order $\hbar$. Therefore, the iterations of Equation 157 produce an absolutely convergent series, where the coefficients are formally independent of $\hbar$, but in general go to zero when $\hbar \to 0$. We can therefore say that we have found a solution of Equation 152 in the limit $\hbar \to 0$, because the first few iterations of Equation 157 produce accurate solutions in this limit. It is possible to obtain such accurate solutions because the matrix elements of Equations 153 and 154 are independent of $\hbar$. However, when they depend on the inverse power of $\hbar$ (e.g., in the case when nuclear degrees of freedom are considered), then Equation 157 is not a convenient solution because iterations produce a series in the inverse powers of $\hbar$, and in the limit $\hbar \to 0$ each coefficient of this series becomes infinite.

Very often, when all analytic tools have finally been exhausted one resorts to solving the set of Equations 144 numerically. It is therefore worth mentioning some of the difficulties which are encountered in this case. The difficulties appear when nuclear degrees of freedom are involved and are primarily caused by the fact that the solutions oscillate rapidly and that there are many channels. As we have seen all this is a consequence of the fact that for atoms and molecules the Planck constant is "small but not zero". There is, however, one particular difficulty which we must discuss in more detail and which may greatly affect the calculation when long-lived states are involved. The numerical approach to solving the molecule collision problem is efficient provided the matrix elements of Equation 153 are small. This is the case of the gradient of V, with respect to **r**, is small, i.e., the potential along the collision coordinate changes slowly. That is why this approach is also called adiabatic. On the other hand, the set of equations may have a great drawback which is not so obvious, but becomes self-evident when long-lived states are formed or when large energy transfers are involved. Because of its hidden nature we will discuss it using a simple model. We will analyze the case of the colinear collision of a particle with mass M with a system of two particles, which are bounded together in a square well. The two bound particles are identical with mass m. The incoming particle interacts with the nearest particle as if it were a hard sphere. In Figure 8 we show this system, together with the relevant coordinates.

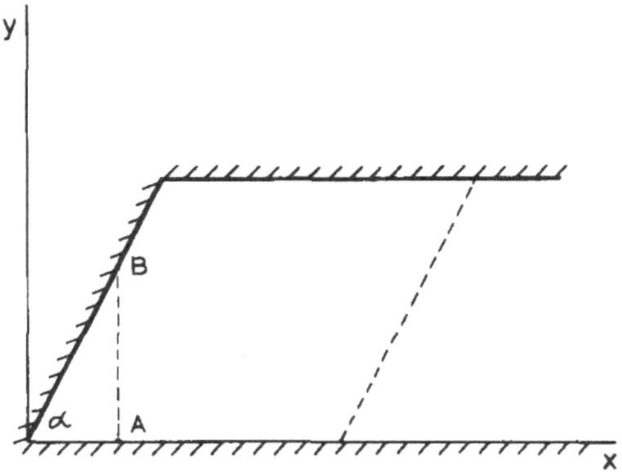

FIGURE 9. Contour of potential for the collinear system in Figure 8 in mass rescaled coordinates

This is the simplest energy transfer collision problem and can be solved by techniques which are different from the straightforward solution of the multichannel equations.[19,20] Since those techniques are only applicable to this example, we will not use them since our purpose is to show how the adiabatic Equations 152 treat this problem.

The Schrödinger equation for this system is

$$\frac{1}{M}\frac{\partial^2 \psi}{\partial x_1^2} + \frac{1}{m}\left(\frac{\partial^2 \psi}{\partial x_2^2} + \frac{\partial^2 \psi}{\partial x_3^2}\right) = [V(x_2 - x_1;\ \ x_3 - x_2) - E]\,\psi \tag{158}$$

where x_1, x_2, and x_3 are the coordinates of particles in the L system. When the coordinates are transformed into the C system and they are properly mass rescaled, Equation 158 becomes

$$\frac{\partial^2 \psi}{\partial x^2} + \frac{\partial^2 \psi}{\partial y^2} = [V(x, y) - k^2]\,\psi \tag{159}$$

where x and y are the new coordinates (note that they are related to those in Figure 8).

Formally the problem has been reduced to a single particle collision in two dimensions on the potential V(x, y), which is shown in Figure 9, where the angle α is determined by the masses M and m[20]. The particle, which comes from the right, oscillates between the two parallel walls and hits the side wall. When it is reflected from this wall it leaves in the direction to the right, but with a different oscillation frequency. The horizontal velocity component of the particle in Figure 9 represents the incoming particle in Figure 8, and its oscillatory movement represents the oscillator in Figure 8.

Let us now solve this problem by the adiabatic method. The solution of Equation 159 is expanded in the series

$$\psi = \sum_n \varphi_n(y, x)\,\omega_n(x) \tag{160}$$

where φ_n (y, x) are the eigenfunctions of

$$\frac{\partial^2 \varphi_n}{\partial y^2} = [V(x, y) - k_n^2]\,\varphi_n \tag{161}$$

and depend parametrically on x. It is obvious that they are only important in the region of the side wall and that they represent standing waves between the points A and B. The normalized eigenfunctions φ_n are

$$\varphi_n = \sqrt{\frac{2}{x \operatorname{tg} \alpha}} \sin \pi n \frac{y}{x \operatorname{tg} \alpha} \tag{162}$$

and the eigenvalues

$$k_n^2 = \frac{\pi^2 n^2}{x^2 \operatorname{tg}^2 \alpha} \tag{163}$$

$$\omega_m'' + \frac{1}{x} \sum_{n \neq m} \frac{mn}{m^2 - n^2} (-)^{m+n} \omega_n' + \frac{1}{x^2} \sum_{n \neq m} \frac{1}{(m^2 - n^2)^2} \omega_n = \left(k^2 - \frac{\pi^2 m^2}{x^2 \operatorname{tg}^2 \alpha} \cdots\right) \omega_m \tag{164}$$

The coefficients in this set of equations are singular at $x = 0$. We also notice that as $h \rightarrow 0$ the number of channels increases as h^{-1} (the vibrational energy in this particular case is of order h^2 because the potential is not that of a harmonic oscillator). However, in the limit $x \rightarrow 0$ most of the channels are closed though, unfortunately, this does not mean that the prôblem in this region can be reduced to one with fewer channels. In order to see this we should notice that the expansion in Equation 160 is in fact expansion of ψ in a series of standing waves at one particular x value. At a given value of x the maximum number of nodes which the wave function ψ can have is $(1/\pi)\, k\, x \operatorname{tg} \alpha$. For $x = \pi/k \operatorname{tg} \alpha$ this number is equal to one, which means that below this x the amplitude of ψ rapidly goes to zero.

In general, the wave function at such a close distance is not a standing wave of the form $\varphi_0 (y; x)$, and therefore the expansion in Equation 160 has more than one term. This is best demonstrated if we follow the incoming wave as it is scattered from the side wall. For simplicity, it will be assumed that initially the wave travels parallel to the x-axis (this is never realized because of the zero point energy, but as a model it describes the real situation very well). This wave is reflected from the side wall and then travels only in the exceptional case (when $\alpha = 45°$) in the direction perpendicular to the x-axis. In general, after collision it travels to the right and oscillates between the lower and upper boundary. Very often, in particular when the angle α is small, after the first reflection from the side wall the wave travels towards $x = 0$, is scattered again from the side wall, and then exits the region. As we see, the scattering process is quite complex, and it is further complicated if we imagine that the approaching particle interacts with a target particle under the influence of potential which has an additional, e.g., square, well. Then the potential in Figure 9 will have a parallelogram-shaped well, where its right boundary is indicated by a broken line. For certain initial conditions the wave will be trapped inside the wall and will oscillate between the two side walls until it gets out. In such a case the wave function, because of these multiple reflections, is a complicated interference pattern everywhere, and in particular near $x \sim 0$.

Therefore, in general, ψ only resembles a standing wave φ_0 at very small distance and can be written as

$$\psi = f(y) \sin \pi \frac{y}{x_0} \tag{164a}$$

where f(y) is a smooth function. The coefficients of the expansion in Equation 160 are now

$$\omega_n(x_0) = \sqrt{\frac{2}{x_0 \operatorname{tg} \alpha}} \int_0^{x_0 \operatorname{tg}\alpha} dy \sin \pi n \frac{y}{x_0} \sin \pi \frac{y}{x_0} f(y) \tag{165}$$

which in the limit $n \rightarrow \infty$ behave as

$$\omega_n(x_0) \sim \frac{1}{n^3} [f'(x_0 \operatorname{tg} \alpha) - f'(0)] \tag{166}$$

Notice that in this particular example the rate of convergence of the adiabatic expansion is of order n^{-3} in the number of channels. Also, the absolute number of the basis functions depends on the magnitude of the difference between the ideal situation and reality, i.e., how well the basis functions in Equation 162 reflect the true situation. This number depends essentially on the difference between the slopes of the wave function at the two boundaries, and this in turn depends on the physical process of collision. It is evident that for the long-lived states, the difference in Equation 166 is expected to be large when multiple collisions are involved, and therefore the adiabatic expansion becomes progressively more difficult to implement. This also means that the closed channels in Equation 164 have no meaning insofar as forbidden processes in the interaction region are concerned. In fact, it is possible, in principle, to have one open channel for a particular x, but in order to achieve convergence one should include many more closed ones. The question is whether one can use a much better approximation than the adiabatic one so that these difficulties do not arise. There may be cases when this is possible, but they are exceptions rather than a rule, especially when long-lived states are involved and when essentially random reflections occur in the interaction region. No basis set can be adequate for such a situation, and therefore one has to be content with the fact that solving the set of Equations 164 (or some similar set) is a difficult problem.

We will stay longer with this model and discuss another very important problem in the short wavelength scattering limit. At the same time this discussion will be a good introduction to the semiclassical theory which is based on the classical trajectories. Let us look again at how the incoming wave

$$\psi = 2ie^{-ik_x x} \sin k_y y \tag{167}$$

behaves as it approaches the side wall in Figure 9. We consider a particular point on such a wave and follow it as it travels until it escapes to the right. First of all we notice that k_y must have the value $\pi n/a$ in order that ψ satisfies the boundary conditions at the lower and upper walls in Figure 9. For the moment, however, we will write it as a general k_y. The plane wave (Equation 167), in fact, represents two waves which travel to the left. When $\sin k_y y$ is written as the sum of two exponentials, ψ is

$$\psi = e^{-ik_x x + ik_y y} - e^{-ik_x x - ik_y y} = \psi_I - \psi_{II} \tag{168}$$

These two waves travel in different directions, and the best way to illustrate this is to fix their wave fronts and show that the perpendicular rays to them have different angles with respect to the x-axis. The wave front of these two waves is defined as the line along which the phases of Equation 168 are constant. Let us designate this constant by $-k_x C$, in which case the wave front of ψ is defined by equations

$$\begin{aligned} -k_x x + k_y y &= -k_x C \\ -k_x x - k_y y &= -k_x C \end{aligned} \tag{169}$$

which represent two lines, shown in Figure 10a.

The perpendicular rays to these two lines describe wave propagation and, as shown in Figure 10a, they have different angles with respect to the x-axis. Let us now fix a point on

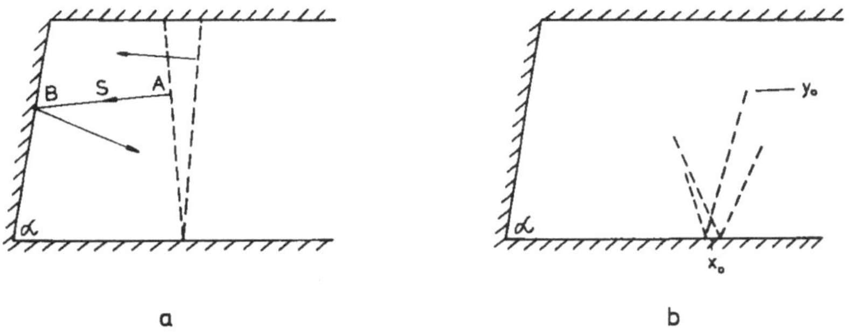

FIGURE 10. (a) Wave front of the incident particle Arrows indicate direction of propagation of the wave front. (b) Scattered wave front.

one of the wave fronts, say A in Figure 10a. It will propagate along a straight line S and will reach the side wall at some point B. At this point the plane wave must be zero, and in order to satisfy this condition we must introduce the reflected wave, i.e., we must write

$$\psi_{II} = e^{-ik_x x - ik_y y} + a\, e^{ik'_x x + ik'_y y} \tag{170}$$

where k'_x and k'_y are the wave numbers which are chosen in such a way that ψ_{II} is zero along the line $y = x \text{ tg } \alpha$ (i.e., along the side wall), i.e.,

$$\psi_{II} = e^{-ik_x x - ik_y x \text{tg}\alpha} + a\, e^{ik'_x x + ik'_y x \text{tg}\alpha} = 0 \tag{171}$$

This condition can only be satisfied if $a = -1$ and

$$-k_x - k_y \text{ tg } \alpha = k'_x + k'_y \text{ tg } \alpha \tag{172}$$

or if we write $\text{tg } \phi = k'_y/k'_x$ and $\text{tg } \phi_0 = k_y/k_x$, then

$$-\cos(\phi_0 - \alpha) = \cos(\phi - \alpha) \tag{173}$$

for which the only acceptable solution is

$$\phi = \pi - \phi_0 + 2\alpha \tag{174}$$

This is exactly the angle to be expected if the wave were reflected from a mirror. Therefore, when the point A arrives at the wall, it is reflected according to the standard laws of geometric optics and travels in the direction indicated in Figure 10a. At any subsequent encounter with any of the boundaries the point A is, of course, reflected according to the same laws, hence we can follow its path all the way to the exit. The same reasoning applies to any point on the wave front so that after sufficient time has elapsed we can reconstruct the scattered wave front by joining all the points by a line.

The appearance of the wave front (Figure 10a) after it has scattered from the side wall is shown in Figure 10b. We notice several important points. First, there are *two* wave fronts, one of which lags behind the other. They have different values of k_x and therefore travel at different speeds. The fact that there are two waves means that the parts of the initial wave were scattered by the side wall at two impact angles and therefore were reflected into two different final states. Physically, this means that when the incoming particle hits the oscillator for two different phases (either when contracting or expanding) the energy transfer involved in each case differs. An even more important result is that for a given x, e.g., x_0 in Figure 10b, the wave fronts do not cover the whole range between the two walls. This means that the scattered wave function ψ is only nonzero up to the point y_0 and zero beyond. Strictly speaking, this is not quite true since ψ and ψ' must be continuous everywhere because ψ is the solution of the Schrödinger equation. Therefore, in reality the solution, which we show in Figure 10b, is nonzero up to the point y_0 and when rapidly vanishes (in the short wavelength limit) beyond this point. This region is called the diffraction region and cannot be described by classical theory. It is entirely a manifestation of the wave nature of particles.

Therefore, the scattered wave function is Figure 10b,

$$\psi = \theta_1 e^{ik_x'x + ik_y'y + i\phi_1} - \theta_2 e^{ik_x'x - ik_y'y + i\phi_1} + \theta_3 e^{ik_x''x + ik_y''y + i\phi_2} - \theta_4 e^{ik_x''x - ik_y''y + i\phi_2} \tag{175}$$

where θ_i is zero beyond a certain value of y, but otherwise it is equal to 1. The function in Equation 175 represents the scattered wave ψ. It should be emphasized that both k'_y and k'' do not, in general, satisfy the quantization condition, as was required for the incoming wave. These wave numbers can have any value, and are primarily determined by α. In order to obtain the observed states, however, the scattered wave function (Equation 175) should be expanded in the series of the eigenstates of the oscillator. Thus, we write

$$\psi = \sum_n f_{n,n_0} e^{ik_nx} \sin\left(\pi n \frac{y}{a}\right) \tag{176}$$

where n_0 is the quantum number of the initial state (Equation 168). From Equation 176 we obtain f_{n,n_0}, but first k_x must be written as

$$\hat{k}_x = \left(k^2 - \frac{\partial^2}{\partial y^2}\right)^{1/2} \tag{177}$$

so that ψ is

$$\psi = e^{i\hat{k}_xx}(\theta_1 e^{ik_y'y + i\phi_1} - \theta_2 e^{-ik_y'y + i\phi_1} + \theta_3 e^{ik_y''y + i\phi_2} - \theta_4 e^{-ik_y''y - i\phi_2}) \tag{178}$$

The function in the parentheses can now be expanded in the series in Equation 176, and a typical term in f_{n,n_0} is

$$\int_0^{y_1} dy\, e^{ik_y'y} \sin\left(\pi n \frac{y}{a}\right) = \frac{e^{iy_1\left(k_y' - \pi\frac{n}{a}\right)} - 1}{2\left(k_y' - \pi\frac{n}{a}\right)} - \frac{e^{iy_1\left(k_y' + \pi\frac{n}{a}\right)} - 1}{2\left(k_y' + \pi\frac{n}{a}\right)} \tag{179}$$

This term has a maximum in the vicinity of $k_y' = \frac{n}{a}$ exactly the value of k_y for the system when the final state is properly quantized. Away from this maximum value the coefficients f_{n,n_0} fall off essentially as n^{-1}, though on average faster because of the oscillatory term in the numerator of Equation 179.

The expansion of the scattered wave function (Equation 178) is now

$$\psi = e^{i\hat{k}_x x} \sum_n f_{n,n_0} \sin\left(\pi n \frac{y}{a}\right) = \sum_n f_{n,n_0} e^{ik_n x} \sin\left(\pi n \frac{y}{a}\right) \tag{180}$$

where $k_n = (k^2 - \pi^2 \frac{n^2}{a^2})^{1/2}$.

Let us now go back to the multichannel Equations 164 which describe the same collision. Usually, they are solved for a finite number of channels, which means that the sum in the expansion in Equation 176 is truncated at some value $n = n_{max}$. The expansion in Equation 176 then looks like

$$\tilde{\psi} = \sum_n^{n_{max}} e^{ik_n x} \sin\left(\pi n \frac{y}{a}\right) f_{n,n_0} \tag{181}$$

The difference between $\tilde{\psi}$ and the exact one is of the order n_{max}^{-1}, which follows from the estimate in Equation 179. Therefore, the appropriate coefficients f_{n,n_0} in Equation 181 also deviate from the exact ones by the same order of magnitude. This means that the expansion of the scattering amplitude, which is based on the multichannel Equations 164 is poorly convergent. This rate should not be confused with the rate of convergence of the solution of Equations 164, because they represent two different things. The solution of the multichannel Equations 164 does not represent the correct solution for scattering. To put it simply, the scattering amplitude is obtained from Equations 164 by multiplying that solution by the inverse of one of the Jost functions. Thus, f_{n,n_0} is essentially given by

$$f_{n,n_0} = \sum_n J_{n,m}(J^-)^{-1}_{m,n_0} \tag{182}$$

where J and J^- are the Jost functions. The sum over the intermediate m states may be poorly convergent, but this has nothing to do with the rate of convergence of the procedure for obtaining J and J^-.

How relevant are our conclusions for a more general case? First of all, we have treated a system with relatively few degrees of freedom; second, the potentials are not so "sharp-edged" as we have assumed; and third, in reality such an extremely short wavelength limit is not often realized. The conclusion we thus arrive at is that basically the problems which we discussed do not disappear, but they are less pronounced. For example, in a potential with smooth edges the scattered wave function (Equation 178) may not have such a sharp cut-off as we have assumed, but its overall appearance may be similar. This also means that in systems involving substantial energy transfer, and when long-lived states are formed, the scattered wavefront may be quite complex and consist of numerous components spread over a large range of x. In three dimensions the wave front should look something like that shown in Figure 3b. In all such cases the expansion in Equation 176 (or the three dimensional analogue) exhibits poor convergence. Therefore, in conclusion we can say that all the problems mentioned are present in realistic systems and in those with more degrees of freedom, but sometimes they are less pronounced.

The treatment of the simple problem which we have discussed can be generalized to the more realistic case of three dimensional scattering involving smooth potentials. This generalized form is also known as the semiclassical theory of scattering and is based on classical trajectories. Because of its importance we shall derive its basic principles and discuss its ability to describe long-lived states.

As we have seen, a very simple model, based on the wave front propagation, gave a reasonably good description of the scattering, and therefore it is of interest to give more general formulation of its principles. An approach along these lines was first used in the formulation of the geometric optics[21] and later in the formulation of the quantum theory.[7,22] In the latter case the theory is also known as the method of the "path integrals".

The basis of the theory is assumption that an initial wave, before it enters the interaction region, can be described by a wave of the form

$$\psi = A\, e^{i\sigma - iEt} \tag{183}$$

where we have used the units with $\hbar = 1$. The amplitude A and the phase σ are functions of coordinates. Very often the initial state of a system is not in the form in Equation 183, but is given as

$$\psi = e^{ikz}\, \varphi_n \tag{184}$$

where φ_n is an eigenstate of isolated system. However, in the short wavelength limit, this state can often be represented as a superposition of two waves which are of the form in Equation 183, e.g., if φ_n is an eigenstate of the harmonic oscillator, then for even n

$$\varphi_n \sim \cos(\sqrt{2n+1}\, x) = \frac{1}{2}\left(e^{ix\sqrt{2n+1}} - e^{-ix\sqrt{2n+1}}\right) \tag{185}$$

and so Equation 184 is indeed a combination of the form in Equation 183. In such a case we consider the two components separately, and for each one of them the wave front is defined as t = 0 by the equation

$$\sigma = C \tag{186}$$

where C is a constant. For any subsequent time the equation becomes $\sigma = C + Et$ and defines the time evolution of the wave front. In this way we could in principle follow the whole scattering process if we knew the relationship between the coordinates of the wave front and the dynamics of the system. In order to find this relationship we will assume that Equation 183 satisfies the Schrödinger equation, which is given in a generalized form

$$(-\nabla^2 + V)\, \psi = E\, \psi \tag{187}$$

where ∇ is the gradient operator (in many dimensions) which also contains the masses of particles. From Equations 183 and 187 one obtains

$$-\Delta A - 2i\nabla A\nabla\sigma - iA\Delta\sigma + A\nabla\sigma\nabla\sigma = (E - V)\, A \tag{188}$$

and if it is assumed that σ is large (which is a consequence of the short wavelength limit) then

$$\nabla\sigma\, \nabla\sigma = (E - V) \tag{189}$$

Let us now define a unit vector $\hat{n}$ perpendicular to the wave frcnt. It is given by

$$\hat{n} = \frac{\nabla\sigma}{|\nabla\sigma|} = \frac{\nabla\sigma}{(E - V)^{1/2}} \tag{190}$$

and a small displacement of the length ds along this line is

$$\mathbf{dr} = \mathrm{ds}\,\hat{\mathrm{n}} = \frac{\nabla\sigma}{(\mathrm{E} - \mathrm{V})^{1/2}}\,\mathrm{ds} \tag{191}$$

This is an equation for the family of rays which are perpendicular to the wave front. This equation can be put into a more convenient form if we take the derivative

$$\frac{\mathrm{d}}{\mathrm{ds}}\left[(\mathrm{E} - \mathrm{V})^{1/2}\frac{\mathrm{d}\hat{\mathrm{r}}}{\mathrm{ds}}\right] = \frac{\mathrm{d}}{\mathrm{ds}}\nabla\sigma \tag{192}$$

and by using identity

$$\frac{\mathrm{d}}{\mathrm{ds}}\nabla\sigma = \frac{1}{(\mathrm{E} - \mathrm{V})^{1/2}}(\nabla\sigma \cdot \nabla)\,\nabla\sigma \tag{193}$$

then

$$\frac{\mathrm{d}^2\,\mathbf{r}}{\mathrm{d}\tau^2} = -\frac{1}{2}\nabla\mathrm{V} \tag{194}$$

where $(\mathrm{E} - \mathrm{V})^{1/2}\,\mathrm{d}\tau = \mathrm{ds}$. Equation 194 resembles the Newton equations of motion except for the factor $^1/_2$, which is present because of the way in which the units were taken. Together with the initial conditions at $\tau = 0$ for **r** and

$$\frac{\mathbf{dr}}{\mathrm{d}\tau} = \nabla\sigma; \quad \tau = 0 \tag{195}$$

Equation 194 reconstructs the wave front at any subsequent "time" τ. In practice the phase σ at "time" τ can be obtained from Equation 191 by evaluating the integral

$$\int_0^\tau \nabla\sigma\,\mathbf{dr} = \sigma(\tau) - \sigma(\tau = 0) = \int \frac{\mathbf{dr}}{\mathrm{d}\tau}\,\mathbf{dr} \tag{196}$$

It is on the basis of Equation 194 that one can say that the classical trajectories describe the motion of the wave and that these two concepts are equivalent. Of course, this conclusion only applies in the short wavelength limit.

However, to obtain the wave function we still need A at "time" τ. The equations for A are obtained from Equation 188 if the term of the order σ is set equal to zero, i.e.,

$$2\nabla\mathrm{A}\nabla\sigma + \mathrm{A}\Delta\sigma = 0 \tag{197}$$

from which it follows that

$$2\,\frac{\mathrm{dA}}{\mathrm{d}\tau} = -\mathrm{A}\Delta\,\sigma \tag{198}$$

The solution is

$$\mathrm{A} = \mathrm{A}(\tau = 0)\,\mathrm{e}^{-\frac{1}{2}\int_0^\tau \Delta\sigma \mathrm{d}\tau} \tag{199}$$

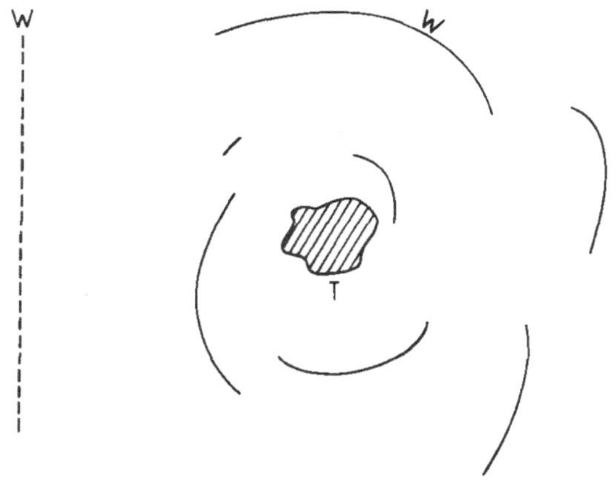

FIGURE 11. Scattered wave front (solid lines) in realistic case, when long-lived states are formed.

There are other ways of expressing A in terms of its initial value $A(\tau = 0)$, but for our purpose it is sufficient to know that the knowledge of the initial wave front σ is sufficient to reproduce the wave function at any other point in time. In general, the wave front after collision will be some very complicated curve. We have already given an example in Figure 10b of how two separate wave fronts appear after collision, which only connect through the diffraction region. In fact, there may be more complicated cases where a single wave front appears after collision as a sequence of the wave fronts, which do not even exist at the same time. The argument for this is in the theory just described. Any single point on a wave front has its own trajectory, and therefore its position is determined by the behavior of the trajectory. If several trajectories from a wave front describe a very short-lived collision, then the appropriate part of the wave front will exit quickly. On the other hand, if several other trajectories involve long-lived states, then the corresponding segment of the wave front will lag behind the former by a significant distance. Therefore, the wave front after collision may appear as shown in Figure 11, where the initial wave front is depicted by a dotted line.

At a particular point, far away from the interaction region, the wave function for stationary collision is a linear combination of all those parts of the scattered wave front which pass through this point. Of course, they do not have the same value of the phase σ. Therefore, the wave function will be a complicated function of the coordinates of all the particles and, going from one scattering angle to another, it will change almost randomly. From such a wave function one obtains the scattering amplitudes by expanding ψ into those asymptotic states which correspond to the boundary condition. It is obvious that the resulting scattering amplitudes may be quite complicated functions of the quantum numbers even though the structures of the scattered wave fronts may be quite simple.

This theory has a serious limitation; it fails in the case when the width of the wave front is very small. By small we mean that when the width is comparable to the local wavelength. It is obvious that in such a case the spread of the wave is large, and after a while the narrow wave front will so much diffuse that it will no longer exist. It has effectively been dispersed throughout the surroundings and is propagating as a spherical wave. These narrow wave fronts are encountered when long-lived states are formed in the unstable systems. In such systems any point in the interaction region which acts as a source of exponential separation contributes to the breakdown of the theory, and the longer the trajectory stays within this region the less well it will represent the propagation of a real wave front. After a while the importance of such a wave front becomes academic, which means that the narrow wave fronts in Figure 11 will effectively not appear.

There are other limitations to this theory, but they are less severe. For example, the theory has difficulty in explaining resonance or transitions between the electronic states. In general, these are the processes where the classical path does not connect initial and final states, and for this the concept of tunneling of trajectories must be introduced. Furthermore, as we have seen, resonances are not strictly long-lived states, which can be represented by "trapped" trajectories, and therefore tunneling of classical trajectories, which we described in Chapter 2, is only a qualitative way of representing them.

Chapter 4

DECAYS

A special type of evolution in time is decay. The initial conditions which determine decay are different from the scattering problems, and because of this it cannot be treated as the subject of the scattering theory. Since the mathematical description is similar in both cases, because unbound states are involved, the two processes are often treated together. A specific feature of decay is that initially the system consists of a stable molecule, or cluster of particles. At a certain moment, e.g., $t = 0$, the system is destabilized by introducing enough energy into it that after some time it splits into fragments. In the simplest case the fragments are two molecules. The process which was initiated at $t = 0$ and completed when the products are observed is called the decay of the system. The question is, what are the properties of decay?

There are two basic differences between decay and scattering, in addition to the fact that in the former only one molecule is present initially and two, in the latter. The first difference is that the decay process is symmetric with respect to the time reversal, i.e., when the system is destabilized at $T = 0$ decay backward or forward in time is not distinguished. This is not the case in scattering because the fact that *two* molecules approach each other initially makes this problem asymmetric to time reversal.

The second difference is that the initial state of a decaying system is specified by its total angular momentum, which is not the case for scattering. As a result of this, in a scattering problem it is necessary to carry out a partial wave decomposition (or parametrization along similar lines), while in the case of a decay there is only one such amplitude and no decomposition is required.

Therefore, in a general expression for the time evolution of a system, given by

$$\psi = \int d^3 k\ \varphi_{\mathbf{k}}(Q, P)\ a(\mathbf{k})\ e^{-iEt} \tag{200}$$

decay is distinguished from the other processes by applying the above restrictions to $\varphi_{\mathbf{k}}(Q)$. This function is a solution of the stationary Schrödinger equation for n particles (atoms), where Q symbolizes the set of coordinates for the system and P, together with $\mathbf{k}$, the appropriate conjugate momenta. The time reversal symmetry implies that

$$\varphi_{\mathbf{k}}^{*}(Q, P) = \varphi_{\mathbf{k}}(Q, P) \tag{201}$$

while the requirement that $\varphi_{\mathbf{k}}(Q, P)$ be an eigenfunction of the total angular momentum operator implies that there are only spherical waves in the coordinates in which the system decays.

In our discussion we will only consider decays which have two molecules, with only one possible arrangement of atoms as end products. In other words, it will be assumed that a molecule X decays into two molecules A and B, with no possibility that other products are formed. This type of system has been discussed earlier where it was shown that we can describe its evolution using only one set of coordinates, i.e., the set which decomposes into the subset for A, the subset for B and $\mathbf{r}$, the coordinates for the relative separation of A and B. In these coordinates we let $\mathbf{r}$ increase to a large value, in which case the Hamiltonian of the system parametrizes as

$$H \sim H_A + H_B + T_{AB} \tag{202}$$

where T_{AB} is the relative kinetic energy of the two molecules. This means that the system can be described by three noninteracting Hamiltonians, and the motions of the molecules A and B are independent.

If the molecule A is in an eigenstate specified by the set of quantum numbers n_A and total angular momentum j_A, then its eigenfunction can be written as a product $Y_{JA}^{m_A}\omega_{nA}^{A}$ (e.g., in generalized hyperspherical coordinates). Similarly, we can write the eigenfunction for the molecule B. It should be pointed out, though, that the exact parametrization of the eigenfunctions for the molecules A and B is not essential for our derivation of the laws of decay. It is, however, convenient to write these functions in a form which explicitly highlights the essential properties of the decay process.

Therefore, the wave function of the system has a general asymptotic behavior for $r \rightarrow \infty$:

$$\varphi_{\mathbf{k}}^{(n_0)} \sim \frac{1}{r} \sum_{n}^{N} Y_J^M \, \omega_{n_A}^A \omega_{n_B}^B [e^{ik_n r} f_{n,n_0}(k) + e^{-ik_n r} g_{n,n_0}(k)] \tag{203}$$

where the summation index n takes all relevant values which uniquely specify the system. The eigenfunction of the total angular momentum operator is Y_J^M, which contains j_A, j_B, and l, the orbital angular momentum associated with the relative motion of the molecules A and B. These indexes are also contained in n. The functions of f_{n,n_0} and g_{n,n_0} are the so-called Jost functions, which depend only on the modulus of **k**. From the time reversal invariance in Equation 201 it follows that

$$f_{n,n_0}^*(k) = g_{n,n_0}(k) \tag{204}$$

For an N-channel problem (Equation 203) there are N linearly independent solutions (if only open channels are considered), and therefore we have used the index n_0 to indicate each of these solutions. It follows that the function φ in Equation 200 is a linear combination of the form

$$\psi = \sum_{n_0} a_{n_0} \varphi_{\mathbf{k}}^{(n_0)} \tag{205}$$

The decay of the system is now described by

$$\psi = \int d^3 k \sum_{n_0} a_{n_0} \varphi_{\mathbf{k}}^{(n_0)} e^{-iEt} \tag{206}$$

where the amplitude a(k) in Equation 200 was merged with the coefficients a_{n_0}. The coefficients a_{n_0} are obtained from the value of ψ at $t = 0$, which is designated by ψ_0. From Equation 206 we obtain

$$\int \psi_0 \varphi_{\mathbf{k}'}^{(m_0)} dV_A \, dV_B \, dV = \int d^3 k \sum_{n_0} a_{n_0} \int \varphi_{\mathbf{k}'}^{(m_0)} \varphi_{\mathbf{k}}^{(m_0)} dV_A \, dV_B \, dV \tag{207}$$

where the volume element $dV_A \, dV_B \, dV$ designates integration over all degrees of freedom. In order to obtain a_{n_0} we have to calculate the integral

$$I_{\mathbf{k}',\mathbf{k}}^{(m_0,n_0)} = \int \varphi_{\mathbf{k}'}^{(m_0)} \varphi_{\mathbf{k}}^{(m_0)} dV_A \, dV_B \, dV \tag{208}$$

Calculation of this integral is shown in Appendix A, except that $\psi_{\mathbf{k}}^{m_0}$ is now more complicated and has different asymptotic behavior than for scattering. We start from the Schrödinger equation for $\psi_k^{m_0}$

$$H\,\varphi_k^{(n_0)} = (T_{AB} + T_A + T_B + V)\,\varphi_k^{(n_0)} = E\,\varphi_k^{(n_0)} \tag{209}$$

where V is the potential of the whole system. From Equation 209 we obtain

$$(E - E')\,I_{k',k}^{(m_0,n_0)} = \int [\varphi_{k'}^{(m_0)}(T_{AB} + T_A + T_B)\,\varphi_k^{(n_0)} - \varphi_k^{(n_0)}(T_{AB} + T_A + T_B)\,\varphi_{k'}^{(m_0)}] \cdot dV_A\,dV_B\,dV \tag{210}$$

It can be shown that the volume integrals which include T_A and T_B are zero, and therefore the integral in Equation 210 is

$$(k'^2 - k^2)\,I_{k',k}^{(m_0,n_0)} = \int [\varphi_{k'}^{(m_0)}\,\Delta\,\varphi_k^{(n_0)} - \varphi_k^{(n_0)}\,\Delta\,\varphi_{k'}^{(m_0)}]\,dV_A\,dV_B\,dV \tag{211}$$

where $k^2 = 2\mu_{AB}E$. The volume integral in r can be transformed into a surface integral and

$$(k'^2 - k^2)\,I_{k',k}^{(m_0,n_0)} = r^2 \int [\varphi_{k'}^{(m_0)}\,\frac{1}{r}\,\varphi_k^{(n_0)} - \varphi_k^{(n_0)}\,\frac{1}{r}\,\varphi_{k'}^{(m_0)}]\,dV_A\,dV_B\,d\Omega \tag{212}$$

where $d\Omega$ is the solid angle. Since the limit $r \to \infty$ is assumed the function $\varphi_k^{(n_0)}$ can be replaced by its asymptotic form (Equation 203), in which case integration over the solid angle $d\Omega$ and over the volume elements can be done without difficulty. We have

$$(k'^2 - k^2)\,I_{k',k}^{(m_0,n_0)} = \sum_n [e^{ik'_n r} f_{n,m_0}(k') + e^{-ik'_n r} g_{n,m_0}(k')] \cdot \frac{\partial}{\partial r}[e^{ik_n r} f_{n,n_0}(k) + e^{-ik_n r} g_{n,n_0}(k) - (k \rightleftarrows k')] \tag{213}$$

where we have neglected the terms of the order r^{-1}. The symbol $(k \rightleftarrows k')$ means that the preceding term is repeated, with k and k′ switched. When the derivative is calculated and like terms collected, we obtain

$$(k'^2 - k^2)\,I_{k',k}^{(m_0,n_0)} = i \sum_n \{(k_n - k'_n)[f_{n,n_0}(k)\,f_{n,m_0}(k')\,e^{ir(k_n + k'_n)} - g_{n,n_0}(k)\,g_{n,m_0}(k')\,e^{-ir(k_n + k'_n)}] + (k_n + k'_n)[f_{n,n_0}(k)\,g_{n,m_0}(k') \cdot e^{ir(k_n - k'_n)} - f_{n,m_0}(k')\,g_{n,n_0}(k)\,e^{-ir(k_n - k'_n)}]\} \tag{214}$$

In the limit $r \to \infty$ the first term is zero since k_n and k'_n are positive, while the second term gives

$$I_{k',k}^{(m_0,n_0)} = \sum_n \frac{\sin r(k'_n - k_n)}{k'_n - k_n}[f_{n,n_0}(k)\,g_{n,m_0}(k') + f_{n,m_0}(k') \cdot g_{n,n_0}(k)] + \sum_n \frac{\cos r(k'_n - k_n)}{k'_n - k_n}[f_{n,n_0}(k)\,g_{n,m_0}(k') - f_{n,m_0}(k')\,g_{n,n_0}(k)] \tag{215}$$

It can be shown that the last term is zero in the limit $k' \rightarrow k$. This is a consequence of the fact that for the multichannel equations

$$\sum_n k_n f_{n,n_0}(k)\, g_{n,m_0}(k) = \sum_n k_n g_{n,n_0}(k)\, f_{n,m_0}(k) \tag{216}$$

and that the limit $r \rightarrow \infty$ is assumed. Therefore, the integral in Equation 208 is

$$I_{k',k}^{(m_0,n_0)} = \pi \Sigma\, \delta(k'_n - k_n)(f_{n,n_0}g_{n,m_0} + f_{n,m_0}g_{n,n_0}) \tag{217}$$

of if one notices that $k_n = (k^2 - \epsilon_n)^{1/2}$ (where ϵ_n is a constant which corresponds to the threshold energy of the nth channel) then

$$I_{k',k}^{(m_0,n_0)} = 2\,\pi \frac{\delta(k' - k)}{k} \Sigma\, k_n\, f_{n,n_0}\, g_{n,m_0} \tag{218}$$

where Equation 216 has been used.

From this we obtain the amplitudes a_{n_0}, which are solutions of the equation

$$2(2\pi)^2\, k\, \Sigma\, I_k^{(m_0,n_0)}\, a_{n_0} = \int \psi_0\, \varphi_k^{(m_0)}\, dV_A\, dV_B\, dV \tag{219}$$

where $I_k^{(m_0,n_0)}$ designates the sum 218 without the factor in front of it. In matrix form this has the solution

$$a = \frac{1}{2k(2\pi)^2} I_k^{-1} \int \psi_0\, \varphi_k\, dV_A\, dV_B\, dV \tag{220}$$

and the time evolution of the system is

$$\psi = \frac{1}{2\pi} \int dk\, k\, e^{-iEt}\, \varphi_k I_k^{-1} \int \psi_0\, \varphi_k\, dV_A\, dV_B\, dV \tag{221}$$

where φ_k is a column matrix with the components $\varphi_k^{(n_0)}$.

Far away from the interaction region where, in fact, all the observation is done, we can use the asymptotic form in Equation 203 in order to obtain the time evolution of the decaying state. First, let us define some simplified notation. We will write the Jost functions f_{n,n_0} and g_{n,n_0} as the matrices f and g, respectively; the channel vectors k_n, as a diagonal matrix K; and the product $Y_J^M \omega_{nA}^A \omega_{nB}^B$ as a column vector ω_J^M. In this notation the asymptotic form of ψ is

$$\psi \sim \frac{\tilde{\omega}_J^M}{2\pi r} \int dk\, k\, e^{-iEt}\, K^{-1}(e^{iKr}g^{-1} + e^{-iKr}\, f^{-1})\, \psi_0 \varphi_k\, dV_A dV_B\, dV \tag{222}$$

The time analysis of ψ is analogous to that which we have made in the scattering case. There is a difference, however, that in the scattering case one can make a plausible guess for the initial wave packet ψ_0 and calculate its overlap integral with φ_k, where as here this is not possible, though one can assume that the integral is a smooth function of k. This model should be treated with great caution, and we will only take it as an example to demonstrate those features of ψ which are independent of ψ_0.

Let us consider the integral which involves $\bar{g}^{-1}$. Most of the contribution to this integral comes from the stationary point of the phase

$$\delta = -\frac{k^2 t}{2\mu_{AB}} + k_n r + \arg(\tilde{g}^{-1})_{n,n_0} \tag{223}$$

which gives

$$-\frac{kt}{\mu_{AB}} + \frac{k}{k_n} r + \frac{d}{dk} \arg(\tilde{g}^{-1})_{n,n_0} = 0 \tag{224}$$

If we now define the time delay for the component of the decaying state with the wave number k, as

$$\tau = \frac{\mu_{AB}}{k}\frac{d}{dk} \arg(\tilde{g}^{-1})_{n,n_0} \tag{225}$$

then from Equation 224 we obtain the time of arrival of this component at the detector. However, for decay it is not appropriate to ask the arrival time of one component of the decaying state, but instead to look at the time scan of ψ. In scattering it was just the opposite because the initial wave packet ψ_0 has a wave carrier of the wave number k_0, and it is natural to ask the arrival time of this particular component. For a decaying state there is no obviously unique vector, and any k is as good as any other. Therefore, we solve Equation 224 for k when t is arbitrary and r is fixed. If this value is k_s then the component of ψ with $(\tilde{g}^{-1})_{n,n_0}$ is proportional to

$$\psi \sim e^{-i\frac{k_s^2}{2\mu}t + ik_n r}(\tilde{g}^{-1})_{n,n_0} \int \psi_0\, \varphi_{k_s}^{(n_0)}\, dV_A\, dV_B\, dV \tag{226}$$

and therefore its modulus is more or less determined by $(\tilde{g}^{-1})_{n,n_0}$.

Likewise if we calculated the component of ψ with $\tilde{f}^{-1}$ we would have found that the solution of Equation 224 only has an acceptable solution if $t < 0$. Therefore, this component of ψ represents a decaying state going backward in time. In fact, the two components of ψ are the mirror images of each other.

The analysis of ψ with the stationary point is valid if the modulus of $(\tilde{g}^{-1})_{n,n_0}$ does not change rapidly. One obvious case when this is not satisfied is when $\tilde{g}^{-1}$ has a pole in k which is close to the real axis. In such a case analysis must proceed in the same way as for resonances in collisions. The nice thing about the poles of $\tilde{g}^{-1}$ is that they coincide with the poles of the appropriate partial wave scattering amplitude. The reason for this is that the amplitude is proportional to the S-matrix, and the S-matrix is proportional to the product fg^{-1}.

In the vicinity of such a pole, g^{-1} parametrizes as

$$\tilde{g}^{-1} = \frac{G}{k - k_p} \tag{227}$$

where now G is different from the analogous quantity in Equation 117 because, as has been noted, f is not present in decay problems. The analysis with the parametrization in Equation 227 is similar to that in the scattering problem, and therefore at a very short time after $t = 0$, there will be no signal at detector. As time increases the first signal arrives from the components of the decaying state which has a large k value, or very short τ_d. The signal

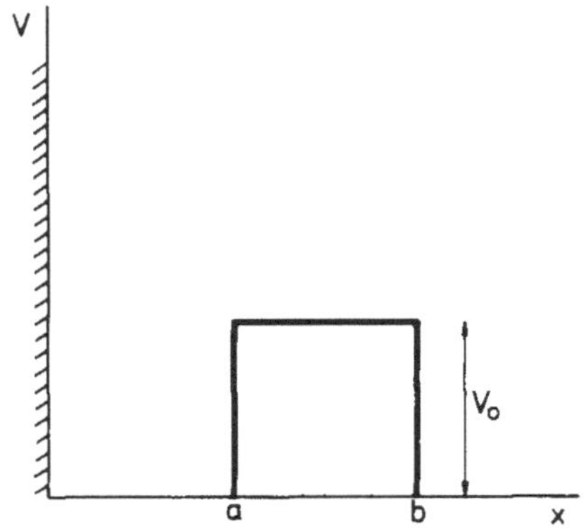

FIGURE 12. Atom-atom potential with a square barrier. In decays the wave function is initially confined to the region inside the barrier.

will continue until the time when $k_s = k_r = \mathrm{Re}(k_p)$. Just before this moment (which is entirely determined by τ_d for $k_s = k_r$) the signal suddenly rises to reach a maximum at about $k = k_r$ and after that decays exponentially to zero. Again we should repeat that the rate of this decay is determined by $\mathrm{Im}(k_p)$. In this analysis, we have neglected the behavior of the overlap integral in Equation 222. When it is included in the analysis the time of arrival of the sudden rise of the wave function may change considerably.

The difference between the time evolution of a decaying system and the time evolution in scattering is large. For a start, in the case of decay one partial wave is present and the system evolves symmetrically in the past and future, which is not the case in scattering. This difference alone makes the two processes completely separate. Another difference lies in the way the initial conditions are chosen, i.e., in the choice of ψ_0. In decay, ψ_0 is confined to a small region inside potential, and in scattering it is often the plane wave. Also in the time evolution of the wave packet, for the scattering case one has the scattering amplitude instead of g^{-1} in Equation 222. Since the scattering amplitude is related to the S-matrix, and the S-matrix to the product $f\,g^{-1}$, it is obvious that decay and scattering are not related processes, because the factor f in Equation 222 is missing.

This brief discussion implies that decay cannot be treated as a part of a scattering process though this is very often assumed when the long-lived states are involved. Namely, it is quite tempting to assume that when such states are formed in collisions, after a certain time they "forget" their history and therefore decay independently of their mode of formation. We will make a brief discussion of what this model means and show that it should be treated with some caution.

Several statements are implied in this model. First, if decay is treated as a part of scattering (loosely speaking, decay is "one half" of scattering), then the intensity of the resulting wave packet should be approximately equal in both cases. There is no reason for this to be true because the intensity is determined by g^{-1} for the decaying state and by the product fg^{-1} in scattering. Intensities, therefore, are very different, as we will show with a simple example. Let us assume a one dimensional scattering and decay problem, on the potential as shown in Figure 12.

The functions f and g can be found explicitly and are given by

$$f^* = g = \frac{e^{ikb}}{4k\,\kappa}\,(k + i\kappa)(k\cos ka + \kappa\sin ka)\,e^{\kappa(b-a)} - (k - i\kappa)$$

$$\cdot\ (k\cos ka - \kappa\sin ka)\,e^{-\kappa(b-a)} \qquad (228)$$

where $\kappa = (V_0 - k^2)^{1/2}$, since it is assumed that $V_0 > k^2$. When $V_0 < k^2$, κ changes to $|\kappa|/i$. For the scattering problem the relevant quantity is the S-matrix, which is the ratio of f and g, i.e.,

$$S = \frac{f}{g} \sim e^{-2ikb} \frac{k - i\kappa}{k + i\kappa} \tag{229}$$

where $\kappa(b - a) \gg 1$. Therefore, the modulus of S is unity, which is also in the vicinity of the pole k_p. The real part of this pole is approximately given by the solution of

$$k \cos ka + \kappa \sin ka = 0 \tag{230}$$

The scattered wave packet will also have a value of the order of unity. On the other hand the amplitude of the decaying state is proportional to g^{-1}, which is approximately

$$g^{-1} \sim e^{-\kappa(b-a)} \tag{231}$$

Therefore, the amplitude of the decaying state will be negligible for all values of k_s which satisfy $k_s^2 < V_0$, even at the resonance. A new picture of a decaying state emerges from this fact and changes that discussed previously. The amplitude of the decaying wave has an appreciable value until the time when $k_s^2 = V_0$, and it quickly goes to zero after this, no matter whether k_s coincides with k_r or not. This fits logically with the conservation of probability rule since the product of the lifetime and intensity of the outgoing current must be constant. Therefore, the longer the resonance takes to decay, the weaker the outgoing current must be.

From this simple example we conclude that decay is not a part of scattering, in the sense which the model assumed. No matter how long-lived the intermediate state in collisions, it cannot be isolated and treated as a decaying state. To put it in other words, if we look at the decay of a long-lived state, it cannot be assumed that its mode of formation is immaterial.

Of course, the conclusion based on the previous simple example cannot be generalized to more complex systems without taking into account the fact that in more complex systems there are other modes of formation for long-lived states than tunneling. In any case, we showed that resonances are not really long-lived states. The exercise, however, showed that care should be taken when using the model, in which decay is "one half" of scattering, because these two processes are not the same.

Another statement which is implied by the model concerns the lifetimes of the decaying state and the lifetime of the intermediate state in scattering. In scattering the lifetime of a state is associated with the delay time (Equation 92). Therefore, if an intermediate state is formed with a very long lifetime this will manifest itself in a very late arrival of the scattered wave packet. Pictorially, we can represent this process as shown in Figure 13a, where particles A and B collide and C and D come out after a long delay.

If it is now assumed that in the "middle" of this path the long-lived intermediate has "forgotten" how it was formed and that it evolves as a decaying state, as shown in Figure 13b, then its lifetime is determined by Equation 225. The result should be approximately one half of the time delay (Equation 92) according to the model. This, however, is not true: (1) there is a formal difference between the quantities which enter the two definitions of time delay (in one it is essentially the ratio $f \cdot g^{-1}$ and in the other simply g^{-1}); (2) the choice of partial waves as an expansion set for the scattering amplitude is not always the most suitable, as we have discussed earlier. This means that in scattering a long-lived state can be formed which has no well-defined total angular momentum, and therefore it is difficult to treat it as a decaying state, which has very well-defined total angular momentum.

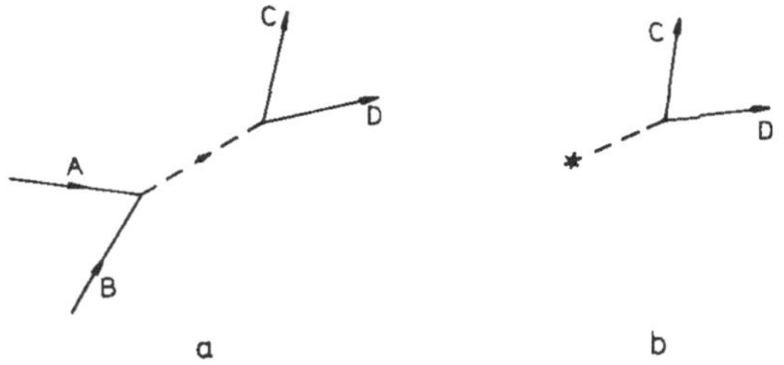

FIGURE 13. (a) Scattering of particles A and B with a long-lived state as intermediary. (b) Decay process of the long-lived state in (a).

There are, however, particular states — resonances — which need more attention since they exhibit rather special behavior. Although they are not true long-lived states, it is rather tempting to formulate the above model for them. The feature of resonances which favors such a model is that they are formed in a particular partial wave, and therefore they can be treated as decaying states because they have well-defined total angular momentum. The problem with a resonance is that its lifetime is not well defined, since it is not a true long-lived state.

When the incoming wave packet arrives at the target, it initiates the formation of a resonance pumping probability current into it. The formation process takes time of the order Γ^{-1}. If the incoming wave packet has, in the meantime, vanished, because it is spatially narrow, the resonance will never be fully formed and it will quickly disappear. Therefore, the lifetime of a resonance, in this case, is arbitrary because it depends on the length of the incoming wave packet. In the extreme case of a very short wave packet the resonance lifetime becomes negligible and, in fact, one cannot talk about the length of time of the resonance since it was never formed. One should say, though, that a signal with a message of this transient process does arrive after a certain time delay, and the value of this time delay is independent of the properties of resonances.

If, on the other hand, the incoming wave packet is spatially long enough, the resonance will be fully formed, in a time of the order Γ^{-1}, and will remain so until it leaves, whereupon the resonance will decay in a time of the order Γ^{-1}. The question is now what one associates with a resonance, which corresponds to the intermediate path in Figure 13a.

First of all, if such a path has any meaning, then the long-lived state should form almost instantaneously so that, in principle, we should be able to observe the event shown in Figure 13a, e.g., in a bubble chamber. It is obvious that for a resonance this is not possible because it takes time to form, and in order to form the initial wave packet must be very broad, i.e., the positions of the particles A and B must be delocalized. In such a case, the event in Figure 13a cannot be observed even in principle. The resonance can only be observed as an effect on the intensity of the scattered wave packet, i.e., indirectly.

There is, however, a period of time during a scattering event after the incident wave packet has gone and the resonance is left when one can consider it as an intermediate state. If it lives long enough in such a case, i.e., Γ is small, it will behave like an intermediate state (Figure 13a). As this part of the scattering event is the same as in the decay problem (as far as the lifetime is concerned), one can say that decay can be treated as part of a scattering process. Indeed, the longer the decay time of resonance, the better this picture is.

The third thing which the model implies is that if a long-lived state is formed in a collision, it will after a certain time "forget" the mode of its formation. This means that the properties of such a decay are more or less independent of ψ_0 in Equation 221. By properties we mean the decay lifetime in a particular channel and the distribution of final states. It appears that this is true for a decay since the overlap integral in Equation 221 is real (both ψ_0 and ψ_k are real), and therefore the phase is zero and thus the time in Equation 225 is unaffected by ψ_0. This, however, may not be the case if the overlap integral is an oscillating function of k, as it has to be written as the sum of two complex components, which are the complex conjugates of each other. The phase of each component then enters the definition in Equation 225 and can affect the decay time. In this way we obtain two components of a decaying state: one with a long decay time and the other with a short one. Of course, only two components appear in the simplest case, but there may be a large (even) number of them in general. Whether the decay time is affected by ψ_0 or not depends on the system, but in general, one can say that it will be especially affected in more complex systems.

The previous conclusions also apply to the distribution of the final states. In short, the answer is that ψ_0 may or may not greatly affect the distribution, which depends on the integral between ψ_0 and ψ_k in Equation 221. In both cases, however, the intuition must not be the guide for such a model; instead one must estimate the overlap integral in Equation 221.

In conclusion, we can say that the model, in which decay is treated as a part of a scattering process, has great inadequacies because the two processes differ and produce different results. There are, however, exceptions, i.e., resonance, but even in this case, the model must be applied in the light of the same stringent assumptions.

The question is how should one prepare a state ψ_0, which is a purely decaying one. By definition ψ_0 is created at time $t = 0$ from a stable state destabilized by introducing enough energy so that it can split into stable fragments. The destabilization must involve some kind of collision with other particles but, as we have seen, in such a case the process is no really a decay but a scattering event. There are other destabilization processes where collisions are indirectly responsible for the creation of decaying states, and in fact sometimes it is impossible to create a decaying state in a direct collision. Consider an α-decaying nucleus with a half-life of 10^6 years, a process known to be caused by a resonance with a very narrow width. This state could not have been created by collision because it would have taken the same length of time to create the state; hardly a realistic picture. Such decaying states are created by indirect mechanism; a particle x is destabilized in a collision with another particle, and it splits into the fragments, e.g., A and B, one of which is again unstable, e.g., A. The behavior of fragment A is described by the laws of collision, but if a resonance is involved, then the appropriate energy component of A behaves after a length of time according to the laws of decay. Creation of a resonance by direct collision is thus avoided; however, this mechanism is rather specific to such systems.

Several conditions must be fulfilled for the creation of a decaying state in a direct collision. First, we have seen that a decaying state has a well-defined total angular momentum. Therefore, if an initially stable particle x, which has total angular momentum J is destabilized by a collision with Y, its total angular momentum must still be J, or some precisely known value. This is only possible if Y is light and it has a well-defined (if possible, small) spin. For a molecule x the perturber Y may be a photon or an electron. In both cases the s-wave approximation is perfectly adequate for the scattering amplitude and the change in J, because the spins of the photon and the electron are known and small. One should be careful, though, when photons are concerned. It is assumed that they change only the electronic energy of the molecule x and not the energy of the nuclear motion. The reason is that when electronic energy changes, it is almost an instantaneous process, compared to changes involving nuclei. Therefore, when the electrons of molecule x absorb a photon they make a transition into a higher electronic state so quickly that the nuclei have hardly moved. This means that the nuclear wave function did not change during the transition whereas the internuclear potential did. In such a case, the nuclear wave function, ψ_0, will represent a very well-defined decaying

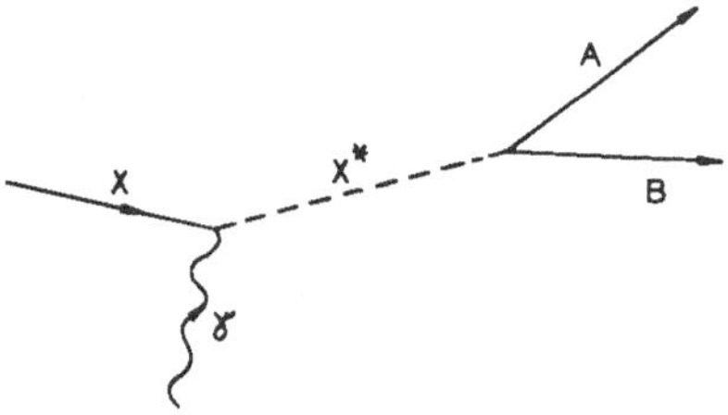

FIGURE 14. Interaction of molecule X with a photon, producing excited state X^+.

state. Although the whole process is essentially a collision of the form shown in Figure 14, there are two favorable aspects which enable us to assume that x* is a decaying state: the change in J for particle x is negligible so that the total angular momentum of x* is well defined and the electron and nuclear motion are almost independent so that the initial nuclear and electronic wave functions are known.

If the interaction of a molecule x with a photon excites the nuclear motion, then the picture of decay is more complicated. First, it should be realized that without knowing the details of the interaction between the photon and the molecule we are unable to obtain ψ_0. Second, if x* is short-lived and breaks up immediately after the interaction with the photon, then there is no reason to treat x* as a decaying state. Therefore, two important points are involved in the study of this kind of decay process which were not present in the previous case.

In order to find ψ_0 one has to solve the equations which describe the interaction of a photon with the molecule x. In the simplest, semiclassical theory this interaction is represented by the electric dipole term, which is time independent. The interaction which starts at $t_1 < 0$ and ends at $t = 0$ has the frequency ω_0 in between. Since the interaction varies in time one has to solve the time-dependent Schrödinger equation with a time-dependent Hamiltonian. The initial condition for the wave function at $t = t_1$ is an eigenstate of x, and the solution ψ_0 at $t = 0$ is the initial condition for a decay process.

In general, the solution ψ_0 depends on many factors, i.e., the strength of the coupling and its duration. It should also be mentioned that in the semiclassical theory of photon-x interaction, spontaneous decays are neglected, and therefore ψ does not reach the steady state, provided x* is much longer lived than t_1, which is *proviso* assumed implicitly. Furthermore, the interacting system of x and the photon, may have resonances which are entirely different from those of the isolated molecule x. In such a case the dominant transitions will be among the nonisolated states, and therefore when x* is left alone it will be predominantly in one of them and not in one of the resonance states of the isolated x*. This means that when the interaction is over, x* is left in a state which, in general, is not an eigen solution of Equation 202.

However, in the case where the photon-molecule interaction is weak we can assume that at $t = 0$ the system is left in an eigenstate of the Hamiltonian for x*, and moreover that it is in one of the resonance states of x. In such a case we can write for ψ_0

$$\psi_0 = \sum_{n_0} \varphi_{k_r}^{(n_0)} C_{n_0} \tag{232}$$

where the basis functions were defined in Equation 203. The coefficients C carry information about how the state ψ_0 was formed and primarily depend on the interaction between the photon and x. The overlap integral in Equation 207 is then

$$\int \psi_0 \varphi_k^{(m_0)} \, dV_A \, dV_B \, dV = \frac{\pi}{k} \delta(k - k_r) \sum_{n_0} C_{n_0} \sum_n k_n [f_{n,m}(k) \, g_{n,n_0}(k_r) + f_{n,n_0}(k_r) \, g_{n,m_0}(k)] \tag{233}$$

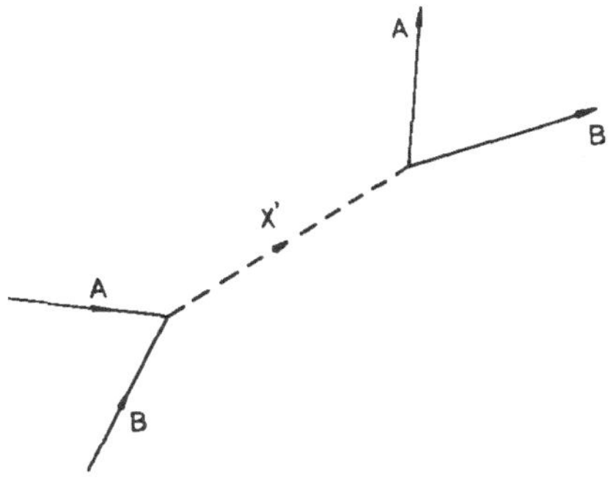

FIGURE 15. Scattering of particles A and B with long-lived intermediary X^+.

where the delta function $\delta(k - k_r)$ must be treated with caution. It is exactly delta function if ψ_0 is exactly the solution of the stationary Schrödinger equation for x. Since ψ_0 is, however, only approximately a stationary solution, $\delta(k - k_r)$ is only approximately a delta function, i.e., it has a maximum at $k = k_r$. The physical reason lies in the uncertainty principle. We know that at $t = 0$ the initial state ψ_0 is localized in the interaction region, i.e., the probability of finding the fragments A and B of x^* is zero outside the doman of x. In such a case Δr is of the order of the radius of the potential which means that the appropriate Δk also has a nonzero value. However, $\delta(k - k_r)$ implies that $k = 0$ which follows from the assumption that $\Delta r = \infty$, i.e., the system is delocalized.

If the overlap integral (Equation 233) is now replaced in Equation 221, and only the component which evolves in the future is retained, then ψ is

$$\psi \sim \int dk\, e^{-iEt}\, \delta(k - k_r)\, e^{ikr}\, K^{-1}\, \tilde{g}^{-1}[\tilde{f}Kg(k_r) + \tilde{g}Kf(k_r)]\, C \tag{234}$$

where we have omitted the nonessential factors and used the matrix notation. The solution is straightforward if $\delta(k - k_r)$ is a delta function, i.e.,

$$\psi \sim e^{-iE_rt + ik_rr}\, f(k_r)\, C \tag{235}$$

which means that it is stationary. This is expected since in this case ψ_0 is an eigenstate of x. If ψ_0 differs slightly from an eigenstate the integral in Equation 234 is more complicated to evaluate. It appears that it has two components, but this is an artifact of the approximation in Equation 232. It is essential, however, to notice that the integrand has a pole in complex k-plane, which comes from g^{-1}, and by assumption it is a resonance pole k_p with a real part k_r. The presence of a pole in the integral inevitably leads to a decaying component of the wave packet, and therefore ψ decreases in time with a rate determined by the imaginary part of k_p.

It is sometimes assumed that the past of the long-lived state x^* is not important and that it forms in a collision as shown in Figure 15 rather than in an interaction between x and a photon. Of course, this model applies only to a single partial wave, and therefore this is not scattering with real initial plane wave.

In such a case coefficients C in Equation 232 can be found and are equal to g^{-1} so that ψ represents the scattering function

$$\psi \sim e^{-ikr} + e^{ikr}\, S \tag{236}$$

where S is the scattering matrix. This model is not realistic because x^* carries the information of its past in the amplitude of ψ and not in the position and the width of resonances. However, the model offers only this.

In conclusion, we can say that there are two aspects of the theory of decay: one is its proper formulation and the other, the preparation of such a state. Strictly speaking, all decay processes are in fact a part of a scattering, but under the same circumstances the details of the latter are not important.

Chapter 5

TWO-ATOM SYSTEMS

I. GENERAL THEORY

A. Transformation of Scattering Amplitude

Typical of atom and molecule collisions is that the electronic motion can be separated from the nuclear motion. For each relative configuration of nuclei one can, therefore, assume that electrons are in an eigenstate of Hamiltonian, in which the kinetic energy of the nuclei is zero. After this separation, one can write the wave function of the entire system in the form of the expansion in Equation 151, in which case the equations of motion for the nuclei are analogous to Equation 152. For two-atom systems the Equations 152, in slightly modified form, are

$$\Delta \psi_m = [V_m(r) - k_m^2]\, \psi_m + \sum_n (\mathbf{v}_{m,n} \nabla\psi_n + \omega_{m,n}\, \psi_n) \tag{237}$$

where now $V_m(r)$ goes to zero for $r \to \infty$. We will call V_m the channel potentials. k_m^2 is the difference $k^2 - E_m$, where E_m is the total energy of separated atoms. The matrix elements $\mathbf{v}_{m,n}$ and $\omega_{m,n}$ are given by Equations 153 and 154, respectively.

The channel potentials do not have a typical shape, except the lowest one, or the ground state potential. Its general shape is shown in Figure 16, but there are cases, although very rare, in which this potential may have a barrier, as shown in Figure 2. However, for the moment the exact details of the channel potentials are of no concern to us, except their limiting values. These potentials are very repulsive for small r, which is due to the exchange effect of electrons and the repulsion of the two nuclei. Repulsion may acquire the value of a few keV for small r, which is much larger than a typical collision energy of atoms, which is of the order of a few eV. Furthermore, the gradient of the potentials at short distances is very large; therefore, we can assume that the nuclear wave functions ψ_m are practically zero for all r which are smaller than some minimum r_0. The value of r_0 is approximately the solution of the equation

$$V_m(r_0) = k_m^2 \tag{238}$$

where we scan all m and take the minimum value of r_0. The meaning of the fact that ψ_m are approximately zero for $r \leq r_0$ is that atom-atom potentials have hard core at $r = r_0$ (or approximately the hard core). This feature of potentials is their most typical characteristic, which must be always emphasized when considering atom and molecule collisions. Other features of $V_m(r)$ are not so general. For example, the tail of the potentials may have analytical form ranging from the Coulombic, i.e, $V(r) \sim r^{-1}$ to $V(r) \sim r^{-6}$, but in fact $V_m(r)$ are never faster decaying than r^{-6}. The channel potentials may have several minima or maxima, and therefore the one in Figure 16 is not typical of $V_m(r)$, other than the ground state.

At this point we should briefly mention that the set of Equations 237 is also encountered in nuclear and elementary particle physics, where it is primarily used for modeling of reactions.[23-25] However, the difference with the set (Equations 237) which describes atom-atom collisions is that with each channel one associates a different mass, i.e., in front of $\Delta\psi_m$ there is a factor (mass of particle in channel m)$^{-1}$, and usually the channel potentials do not have a hard core. Therefore, the theory which we are going to describe can also be

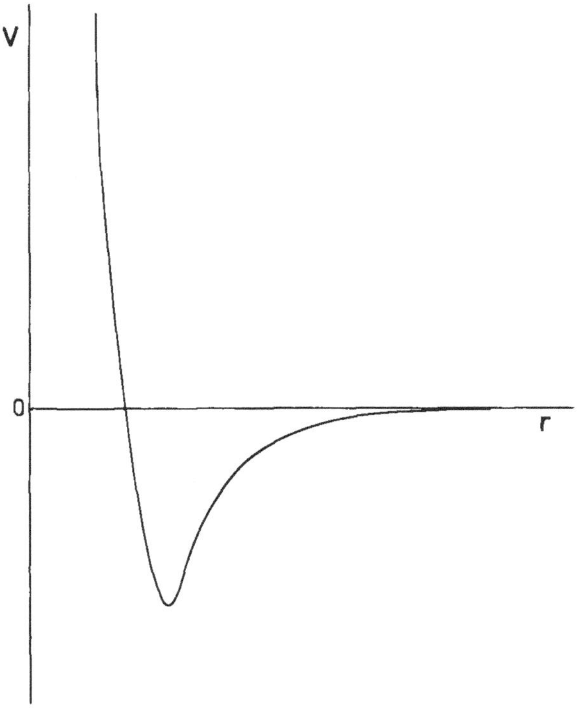

FIGURE 16. Typical ground state atom-atom potential.

applied to such a set of equations, but with the modifications which take into account those differences.

The scattering amplitude is defined by the boundary condition for ψ_m

$$\psi_m(r) \sim e^{ik_m z} + \frac{1}{r}\sum_{n}^{N} e^{ik_n r} f_{n,m}(\Omega) \tag{239}$$

where it was assumed that the incident atom travels in the positive z-direction and that initially the atoms are in electronical state m. The sum extends over all n for which k_n is real.

Solving Equations 237 with the boundary conditions (Equation 239) and $\psi_m(r_0) = 0$ is most conveniently done by expansion of ψ_m in the series of spherical harmonics. We write

$$\psi_m = \frac{1}{r}\sum_{l,\mu} Y_l^{\mu}(\Omega)\, \varphi_{m;l,\mu}(r) \tag{240}$$

where $\varphi_{m,l,\mu}(r)$ is a function of only the module of r. It should be emphasized that solving Equation 237 by making use of Expansion 240 is not the only way of doing this. The convenience of Expansion 240 is not only based on the assumption that the potentials in Equation 237 have spherical symmetry, but also on the fact that the spherical harmonics $Y_l^{\mu}(\Omega)$ are eigenfunctions of the angular part of the kinetic energy operator and that the multidimensional boundary value problem is reduced to a single dimensional problem. This convenience is lost for potentials which do not have such a symmetry and also in the case when Equation 240 is poorly convergent. We will assume for two-atom systems that all terms in Equation 237 have spherical symmetry, in which case the equations for $\varphi_{m,l,\mu}(r)$ have the form

$$\varphi''_m = \left[V_m + \frac{l(l+1)}{r^2} - k_m^2\right] \varphi_m + \Sigma\, v_{m,n}\, \varphi'_n + \omega_{m,n}\, \varphi_n \tag{241}$$

where we have omitted the index (l, μ) of $\varphi_{m,l,\mu}$. The terms $v_{m,n}$ and $\omega_{m,n}$ are not equal to those in Equation 237, although they are designated with the same letters. The set of Equations 241, together with the boundary conditions on φ_m

$$\varphi_{m,n}(r_0) = 0$$

$$\varphi_{m,n}(r) \sim e^{-ik_n r}\, \delta_{m,n} + \sqrt{\frac{k_n}{k_m}}\, S^{(l)}_{m,n}\, e^{-i\pi(l+1)+ik_m r}; \quad r \to \infty \tag{242}$$

replaces the set of Equations 237 with the boundary condition 239. The elements $S^{(l)}_{m,n}$ in Equation 242 constitute a matrix, which is also called the S-matrix (from the scattering matrix). The S-matrix is defined in such a way that for large l it becomes unit matrix. The additional index n of $\varphi_{m,n}$ indicates that the incoming channel is n.

The scattering amplitude in Equation 239 is now given as an expansion

$$f_{n,m}(\Omega) = \frac{1}{2ik_m} \sum_{l=0} (2l+1)\, P_l(\cos\theta)\, (S^{(l)}_{n,m} - \delta_{n,m}) \tag{243}$$

where for convenience we have neglected the ϕ-dependence of $f_{n,m}(\Omega)$, since the interaction of two atoms is spherically symmetric. Likewise, we write the Legendre polynomial $P_l(\cos\theta)$ instead of the spherical harmonics. The sum in Equation 243 is poorly convergent because atoms are relatively heavy particles and the interaction is long range. The typical number of terms in Equation 243 is of order of a few hundred, and therefore analysis of the scattering amplitude in the form in Equation 243 is not easy. This we have already mentioned in Chapter 2, where we have also suggested the use of alternative forms of expansion for $f_{n,m}$. One way of finding alternative expansion is by using the methods of transforming poorly convergent into more rapidly convergent series. Among such transformations, the best known is the Poisson's summation formula, which reads[26]

$$\sum_{l=0}^{\infty} a(l) = \sum_{M=-\infty}^{\infty} e^{-iM\pi} \int_0^{\infty} dl\, a(l - 1/2)\, e^{2\pi iMl} \tag{244}$$

where the series in M is often more rapidly convergent than the series in l. However, the integral over l should be calculated, but sometimes it can be evaluated approximately to a high precision.

We can now apply the formula in Equation 244 to Equation 243, in which case the scattering amplitude in the matrix form is

$$f = \frac{1}{i} \sum_{M=-\infty}^{\infty} e^{-iM\pi} \int_0^{\infty} dl\, l\, P_{l-1/2}(\cos\theta)\, [S^{(l-1/2)} - I]\, K^{-1}\, e^{2\pi iMl} \tag{245}$$

where I is the unit matrix and $K_{m,n} = \delta_{m,n} k_m$. It should be pointed out that transformation of Equation 243, based on the Poisson's summation formula, is not the only way of finding alternative representation of the scattering amplitude, but as it turns out, it produces a more convergent series than Equation 243. We will not consider alternative expansions here, since Equation 245 gives a very good description of scattering.

In Equation 245 we can use the symmetry of the S-matrix in order to replace the sum over the negative M to a sum over the positive M. The assumption of hard core of potential implies that the value of the wave functions φ_m at $r = r_0$ is independent of l, and therefore these functions are invariant to the transformation $l \to -l - 1$, which follows from Equation 241. Therefore, from Equation 242 we find that the S-matrix has the symmetry property

$$S^{(-l-1/2)} = S^{(l-1/2)}\, e^{-2i\pi l} \tag{246}$$

By using Equation 246 and the symmetry of the Legendre functions $P_{-l-1} = P_l$, we can replace the sum over the negative M in Equation 245 by a sum over the positive M and obtain

$$\begin{aligned} f = {} & \frac{1}{i}\int_0^\infty d\, l\, l\, P_{l-1/2}(\cos\theta)\,[S^{(l-1/2)} - I]\, K^{-1} \\ & + \frac{1}{i}\sum_{M=1}^{\infty} e^{-i\pi M}\int_0^\infty d\, l\, l\, P_{l-1/2}(\cos\theta)\,[S^{(l-1/2)} - I]\, K^{-1}\, e^{2\pi i M l} \\ & + \frac{1}{i}\sum_{M=1}^{\infty} e^{-i\pi M}\int_0^\infty d\, l\, l\, P_{l-1/2}(\cos\theta)\,[S^{(l-1/2)}\, e^{-2i\pi l} - I] \\ & \cdot K^{-1}\, e^{2\pi i M l} \end{aligned} \tag{247}$$

Transformation of the scattering amplitude into the form in Equation 245 was possible because the integrands do not have a singular point (e.g., a pole) on the positive real l-axis. In the next section we will show that this is indeed true. In such a case, due to the symmetry of Equation 246 the integrands do not have singularities on the whole real axis, and therefore they also will not have poles in a small vicinity of the real axis. Because of this we can slightly shift the integration path in Equation 247 into the upper complex l-plane, without in fact altering the value of the scattering amplitude. We can show easily that each term in the sum over M is of the order $e^{-2\pi M\eta}$, where $\eta = \mathrm{Im}(l)$, and therefore summation and integration can interchange their sequence, when we obtain for Equation 247

$$\begin{aligned} f = {} & \frac{1}{i}\int_{0^+}^\infty d\, l\, l\, P_{l-1/2}(\cos\theta)\,[S^{(l-1/2)} - I]\, K^{-1} \\ & - \frac{1}{2i}\int_{0^+}^\infty d\, l\, l\, P_{l-1/2}(\cos\theta)\,[S^{(l-1/2)} - I]\, K^{-1}\,\frac{e^{i\pi l}}{\cos(\pi l)} \\ & - \frac{1}{2i}\int_{0^+}^{-\infty} d\, l\, l\, P_{l-1/2}(\cos\theta)\,[S^{(l-1/2)}\, e^{-2i\pi l} - I]\, K^{-1}\,\frac{e^{i\pi l}}{\cos(\pi l)} \end{aligned} \tag{248}$$

where the integration lower limit 0^+ indicates the integration path, as shown in Figure 17.

We can evaluate the second and third integrals in Equation 248 by considering the contour integral

$$I = \oint_{L_1} d\, l\, l\, P_{l-1/2}(\cos\theta)\,[S^{(l-1/2)} - I]\,\frac{e^{i\pi l}}{\cos(\pi l)} \tag{249}$$

where the contour L_1 is shown in Figure 17. On the semicircle of L_1 we will show that the

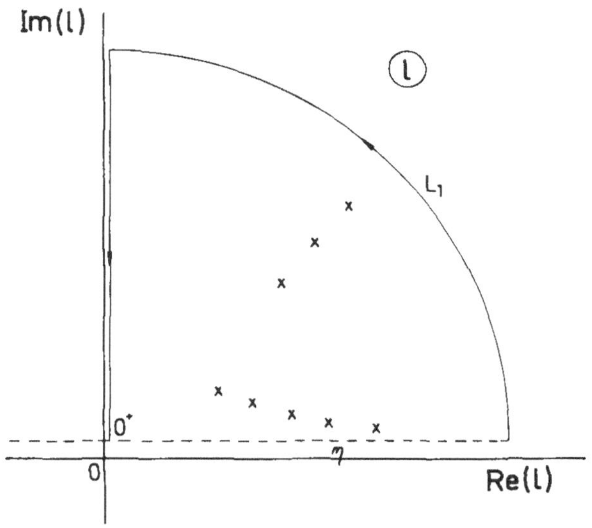

FIGURE 17. Integration contour for evaluation of the integral in Equation 249. Crosses indicate positions of possible poles of the integrand.

integrand of Equation 249 goes to zero. From the asymptotic estimate of $P_{l-1/2}$ $(\cos\theta)$ we have

$$P_{l-1/2}(\cos\theta) \sim O[l^{-1/2}\, e^{\theta \operatorname{Im}(l)}] \tag{250}$$

For l on the semicircle L_1, the centrifugal term in Equation 241 is dominant; therefore, we can write

$$\varphi_n'' \sim \left[\frac{l(l+1)}{r^2} - k_n^2\right]\varphi_m \tag{251}$$

provided $V_m(r)$ goes to zero at $r = 0$ faster than r^{-2}. The hard core at $r = r_0$ is still present for this l; hence, the estimate of the S-matrix is

$$S_{m,n}^{(l-1/2)} \sim -\delta_{m,n}\frac{H_l^{(2)}(k_m r_0)}{H_l^{(1)}(k_m r_0)} \tag{252}$$

where $H_l^{(n)}(x)$ are the Hankel functions. The asymptotic behavior of the Hankel functions for large l is quite complex, but in the region $0 \leqslant \pi/2$ there are essentially two domains in which these functions behave differently.[27] The dividing line of these two domains is defined by the equation

$$\operatorname{Re}(\xi_m) = 0 \tag{253}$$

where

$$\xi_m = (l^2 - k_{mn}^2 r_0^2)^{1/2} - l \ln\left[\frac{l}{k_m r_0} + \frac{(l^2 - k_m^2 r_0^2)^{1/2}}{k_m r_0}\right] \tag{254}$$

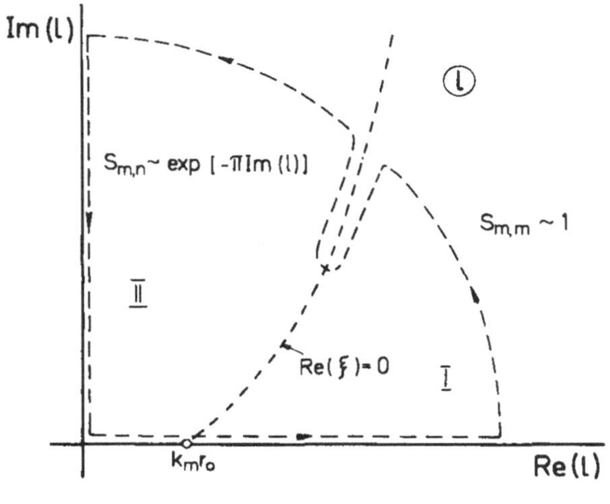

FIGURE 18. Asymptotic behavior of S-matrix in the first quadrant of l-plane.

We show in Figure 18 the relevant region of the l-plane, with the appropriate asymptotic behavior of the S-matrix element $S^{(l-1/2)}_{n,m}$ in the two domains.

Therefore, we can take the worst case for the estimate of $S^{(l-1/2)}_{m,n}$, and that is

$$S^{(l-1/2)}_{m,n} \sim \delta_{m,n}0(1) \tag{255}$$

which we will assume to apply in the whole space $0 \leq \arg(l) \leq \pi/2$.

From this we can now estimate the asymptotic behavior of the integrand of Equation 249 on the semicircle L_1. It goes to zero at least as $l^{1/2} \exp[-(2\pi - \theta)\,\mathrm{Im}(l)]$ so that we can write for I

$$I = \int_{0^+}^{\infty} d\,l\,l\,P_{l-1/2}(\cos\theta)\,[S^{(l-1/2)} - I]\,\frac{e^{i\pi l}}{\cos(\pi l)}$$
$$+ \int_{0}^{\infty} d\,l\,l\,P_{il-1/2}(\cos\theta)\,[S^{(il-1/2)} - I]\,\frac{e^{-\pi l}}{\cos h(\pi l)} \tag{256}$$

On the other hand, the contour integral of Equation 249 is equal to the sum of the residues, if there are poles of the integrand in the domain encircled by L_1. From a look at the integrand we see that they can only be the poles of the S-matrix, since other terms do not have singular points there. Indeed, as shown in the next section, there are poles of the S-matrix in this domain, which we will designate by l_n, and furthermore, they are only first order poles. Very often they are called the Regge poles, after Regge who introduced this technique into the potential scattering.[15] It can also be shown that there are no poles of this kind on the imaginary l-axis. In Figure 17 the Regge poles are indicated by crosses. In general, there is an infinite number of such poles, but the sum of the residues is convergent.[28]

Therefore, we can write for I

$$I = 2\pi i \sum_{\mu} l_\mu P_{l_\mu - 1/2}(\cos\theta)\,\beta_\mu\,\frac{e^{i\pi l_\mu}}{\cos(\pi l_\mu)} \tag{257}$$

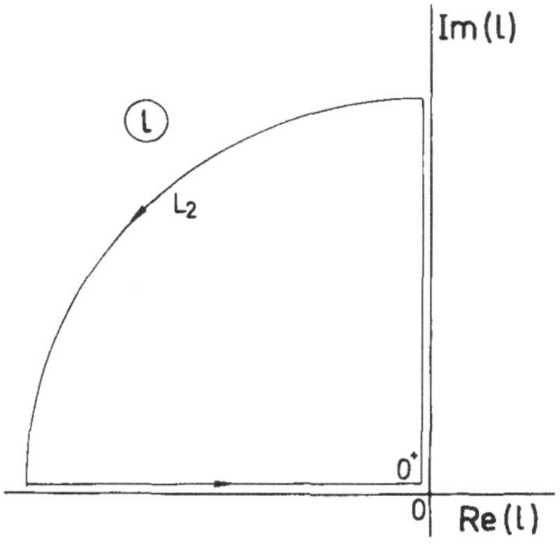

FIGURE 19. Integration contour for the integral in Equation 259.

where

$$\beta_\mu = \lim_{l \to l_\mu}(l - l_\mu)\, S^{(l-1/2)} \tag{258}$$

Similarly, we can evaluate the integral

$$I = \oint_{L_2} d\,l\, l\, P_{l-1/2}(\cos\theta)\,[S^{(l-1/2)}\, e^{-2i\pi l} - I]\, \frac{e^{i\pi l}}{\cos(\pi l)} \tag{259}$$

where now the integration counter is shown in Figure 19.

The asymptotic behavior of $P_{l-1/2}$ (cos θ) on the semicircle of L_2 is the same as in Equation 250. Again the S-matrix elements do not have simple asymptotic behavior, but in the worst case they behave as

$$S^{(l-1/2)}_{m,n} \sim \delta_{m,n}\, 0[e^{-\pi \mathrm{Im}(l)}] \tag{260}$$

and therefore the integrand of I has the estimate $l^{1/2} \exp[-(\pi - \theta)\,\mathrm{Im}(l)]$, which means that the integral on the semicircle of L_2 is zero in the limit $l \to \infty$. The integral I is now

$$I = \int_{-\infty}^{0^+} d\,l\, l\, P_{l-1/2}(\cos\theta)\,[S^{(l-1/2)} e^{-2i\pi l} - I]\, \frac{e^{i\pi l}}{\cos(\pi l)}$$
$$- \int_0^\infty d\,l\, l\, P_{il-1/2}(\cos\theta)\,[S^{(il-1/2)}\, e^{2\pi l} - I]\, \frac{e^{-\pi l}}{\cos h(\pi l)} \tag{261}$$

It will be shown in the next section that there are no Regge poles in the domain $\pi/2 \leqslant \arg(l) \leqslant \pi$; hence, the integral I is equal to zero.

By combining Equations 256, 257, and 261 we find that the scattering amplitude (Equation 248) is

$$f = \frac{1}{i}\int_{0^+}^{\infty} d\,l\,l\,P_{l-1/2}(\cos\theta)\,[S^{(l-1/2)} - I]\,K^{-1}$$
$$+ \frac{1}{i}\int_{0}^{\infty} d\,l\,l\,P_{il-1/2}(\cos\theta)\,S^{(il-1/2)}\,K^{-1}$$
$$- \frac{1}{i}\int_{0}^{\infty} d\,l\,l\,P_{il-1/2}(\cos\theta)\,K^{-1}\frac{e^{-\pi l}}{\cos h(\pi l)}$$
$$- \pi\sum_{\mu} l_{\mu}P_{l_{\mu}-1/2}(\cos\theta)\,\beta_{\mu}\,K^{-1}\frac{e^{i\pi l_{\mu}}}{\cos(\pi l_{\mu})} \tag{262}$$

We can further simplify the expression for the scattering amplitude by using the identity

$$\sum_{l=0}^{\infty}(2l + 1)\,P_l(\cos\theta) = 0; \quad \theta \neq 0 \tag{263}$$

suitably modified by the use of Equation 244. In such a case, for $\theta \neq 0$ we finally have

$$f = \frac{1}{i}\int_{i\infty}^{\infty} d\,l\,l\,P_{l-1/2}(\cos\theta)\,S^{(l-1/2)}\,K^{-1}$$
$$- \pi\sum_{\mu} l_{\mu}\,P_{l_{\mu}-1/2}(\cos\theta)\,\beta_{\mu}\,K^{-1}\frac{e^{i\pi l_{\mu}}}{\cos(\pi l_{\mu})} \tag{264}$$

where the integration path is along the positive imaginary axis and along the positive real axis. This integral can no longer be simplified and must be evaluated by approximate techniques.

In the case when $\theta = 0$ the scattering amplitude must be evaluated from Equation 262. If we use the identity $P_{l-1/2}(l) = 1$, then the first integral in Equation 262 is

$$D = \int_{0}^{\infty} d\,l\,l[S^{(l-1/2)} - I] \tag{265}$$

where now the integration path coincides with the real axis. After partial integration we obtain

$$D = -\frac{1}{2}\int_{0}^{\infty} d\,l\,l^2\frac{d}{dl}S^{(l-1/2)} \tag{266}$$

This integral cannot be evaluated by contour integration because in the region $0 \leq \arg(l) \leq \pi/2$ the integrand does not go to zero for $|l| \to \infty$. It can be evaluated by the stationary phase method. However, we can approximately evaluate the integral by the contour integration in the following way. Let us consider the integral

$$I = \oint_{L_1} d\,l\,l^2\frac{d}{dl}S^{(l-1/2)} \tag{267}$$

where the semicircle of the contour L_1 has a finite but large radius. The integral will have small value on most of the semicircle, provided that in region I of Figure 18 the derivative of the S-matrix goes to zero for $|l| \to \infty$ faster than l^{-3}. However, in the vicinity of the line

$\mathrm{Re}(\xi) = 0$ the derivative of the S-matrix does not go to zero for $|l| \to \infty$. This can be shown from the estimate of the S-matrix elements in Equation 252, near the line $\mathrm{Re}(\xi) = 0$,[27]

$$S_{m,n}^{(l-1/2)} \sim \delta_{m,n} \frac{1}{e^{2\xi_m} - i} \tag{268}$$

so that the integrand of I, for large $|l|$, is asymptotically

$$l^2 \frac{d}{dl} S_{m,n}^{(l-1/2)} \sim \frac{l^2 \ln(l)}{(e^{2\xi m} - 1)^2} \delta_{m,n} \tag{269}$$

If the integration path goes between two poles, then this estimate is $l^2 \ln(l)$. However, Equation 269 only applies in a very narrow region around $\mathrm{Re}(\xi) = 0$; therefore, the overall value of the integral in Equation 267 has somewhat smaller divergency rate on the semicircle L_1. This fact prevents taking the infinite radius of the semicircle of L_1. Nevertheless, we can evaluate the integral in Equation 267 on the finite radius, and in such a case

$$\int_0^\infty d\,l\, l^2 \frac{d}{dl} S^{(l-1/2)} = 4\pi i \sum_\mu^N l_\mu \beta_\mu - \int_\epsilon d\,l\, l^2 \frac{d}{dl} S^{(l-1/2)} - \int_0^\infty d\,l\, l^2 \frac{d}{dl} S^{(l-1/2)} \tag{270}$$

so that for $\theta = 0$ the scattering amplitude is

$$f(0) = -2\pi \sum_\mu^N l_\mu \beta_\mu K^{-1} - \pi \sum_\mu^N l_\mu \beta_\mu K^{-1} \frac{e^{i\pi l_\mu}}{\cos(\pi l_\mu)} + \frac{1}{2i} \int_\epsilon d\,l\, l^2 \frac{d}{dl} S^{(l-1/2)} K^{-1} \tag{271}$$

The index N in the summation indicates the total number of poles encircled by L_1 while ϵ designates integration across the line $\mathrm{Re}(\xi) = 0$. The integral in Equation 262, which does not contain the S-matrix, has been neglected.

The scattering amplitude in Equation 264, together with Equation 271, represents the equivalent form of Equation 243, but is it a more rapidly convergent form? Only after analysis of the Regge poles and residues will we be able to answer this question. At the moment we only notice that f contains two terms. One is a sum over the contribution of the Regge poles and the other is an integral. It may appear that such a form is less convenient than Equation 243 since we do not know anything about the poles and residues. Also the integral seems to be more complicated to evaluate than the sum in Equation 243. However, the integral can be evaluated by approximate methods, and the sum appears to be rapidly convergent since each term has an estimate $\beta_n \exp[-2\pi \mathrm{Im}(l_\mu)]$, which for the Regge poles with large imaginary part is negligible.

Another importance of Equations 264 and 271 is that each term describes certain physical phenomena, while from the form Equation 243 it is more difficult to see how the same phenomena affect the scattering amplitude. We will come to this later, but for the moment we should notice that Equation 264 is not valid in the limit $\theta \to \pi$ because the function $P_{l-1/2}(\cos\theta)$ has singularity at this point. In order to apply similar transformation to f, but the one which will be valid in the vicinity of $\theta = \pi$, we replace $P_l(\cos\theta)$ in Equation 243 by $(-)^l P_l(\cos\theta)$ before making use of Equation 244. In such a case we will have for f

$$f = \sum_{M=-\infty}^{\infty} e^{-iM\pi} \int_0^{\infty} d\,l\,l\,P_{l-1/2}(-\cos\theta)\,[S^{(l-1/2)} - I]\,K^{-1}\,e^{2\pi i Ml + i\pi l} \quad (272)$$

Following the same steps as those in the derivation of the representation in Equation 264 we would find that another form of f is

$$f = -\pi i \sum_{\mu} l_{\mu}\,\beta_{\mu} K^{-1} \frac{P_{l_{\mu}-1/2}(-\cos\theta)}{\cos(\pi l_{\mu})} + \int_0^{\infty} d\,l\,l\,S^{(l-1/2)}\,K^{-1}\,e^{-i\pi l}$$
$$\cdot P_{l-1/2}(-\cos\theta) + \int_0^{\infty} d\,l\,l\,S^{(il-1/2)}\,K^{-1}\,\frac{P_{il-1/2}(-\cos\theta)}{\cos h(\pi l)} \quad (273)$$

Equations 264, 271, and 273 give the complete description of the scattering amplitude in the range of angles $0 \leq \theta \leq \pi$. We will refer to Equations 264 and 271 as the scattering amplitude in the forward space, while to Equation 273 as the scattering amplitude in the backward space. The very fact that there are two representations of f indicates that there are various ways of transforming the scattering amplitude, which we did not explore.*[29] However, those two representations will be perfectly adequate for our purpose, but it would be of interest if they could be superseded by better ones. Historically, the idea of using analytical continuation of the S-matrix into the complex l-plane in order to obtain an alternative form of scattering amplitude was first introduced into the potential scattering theory by Regge.[15] Before that this method was used in the electromagnetic theory, where it was first mentioned by Watson.[14,30] It was later shown to be a powerful tool for description of scattering of electromagnetic waves in the short wavelength limit.[27,31,32] In fact, the most detailed description of the classical rainbow effect was done using this technique.[32] In potential scattering theory the method was introduced via a slightly modified version of summation of series. Instead of using the Poisson's summation formula, the series for f was replaced by the integral[33]

$$f = \frac{1}{2} \oint_c d\,l\,l[S^{(l)} - I]\,K^{-1}\,P_{l-1/2}(-\cos\theta)\,\frac{1}{\cos \pi l} \quad (274)$$

where the contour C encircles the positive real axis. By extending C to include the whole imaginary axis, Regge obtained what is known as the Regge representation of the scattering amplitude

$$f = \frac{1}{2} \int_{-\infty}^{\infty} d\,l\,l\,S^{(il)}\,K^{-1}\,P_{il-1/2}(-\cos\theta)/[\cos h(\pi l)]$$
$$- i\pi \sum_{\mu} l_{\mu}\,\beta_{\mu}\,K^{-1}\,P_{l_{\mu}-1/2}(-\cos\theta)/[\cos(\pi l_{\mu})] \quad (275)$$

It differs from Equation 273 only in the form of integral. This is because in the Regge representation it is implicitly assumed that particle interaction does not have a hard core.[28] If it would have had the hard core the integral in Equation 275 would be divergent. Complex angular momentum technique is a tool for understanding processes in elementary particle physics.[15] Its advantage is in the fact that Regge poles can be associated with resonances, and resonances are essential for many processes in high energy physics. Since resonances

* There is analogous transformation based on Debay's expansion, but it is only useful for potentials with sharp edges.

are also found in collisions of atoms and molecules, electrons and atoms, nuclear collisions, etc. it is believed that this tool may prove to be useful in these fields of research and, in particular, in atom and molecule collisions.

First attempts to apply this technique to atom and molecular collisions were in the original form in Equation 295.[34-36] Very soon it became clear that this form could not account for all processes in these collisions and, in particular, the reflection from the repulsive core. Alternative forms (Equations 264 and 273) were developed,[28] which in fact are proper representations of the scattering amplitude for atom and molecule collisions or collisions in which particles have hard cores. This shows that some of the techniques which were developed for nuclear and high energy collisions cannot be simply transferred to atom and molecule collisions. One reason, as we have seen, is that atom-atom interaction has hard core, but the other is the short wavelength limit of such collisions.

B. Properties of Regge Poles and Residues

Regge poles are defined as poles of the scattering matrix S, which was defined in Equation 242. Such definition of S is not very convenient, principally for two reasons. One is that the wave function φ is very seldom obtained from Equation 241 as a complex solution, which asymptotically behaves as Equation 242. Usually it is obtained as a real solution, which in matrix notation behaves as

$$\varphi^{(l)} = e^{iKr} J^{+}(l) + e^{-iKr} J^{-}(l) \tag{276}$$

The matrices J^{+} and J^{-} are r-independent and are called the Jost matrices or the Jost functions, after Jost who first introduced them in the potential scattering theory.[37] The second reason why the definition in Equation 242 of S is not convenient is that from there it is quite difficult to analyze its analytic structure. Alternative definition of the S-matrix is based on the symptotic form (Equation 276). From there it is quite straightforward to show that

$$S^{(l)} = K^{1/2} J^{+}(J^{-})^{-1} K^{-1/2} e^{i\pi(l+1)} \tag{277}$$

The poles of S^{l} are now the roots of the equation

$$\mathrm{Det}(J^{-}) = 0 \tag{278}$$

in the variable l. J^{+} cannot have poles because the wave function φ, with the boundary condition of Equation 276, cannot be singular. Therefore, by defining S as in Equation 277 we have replaced the problem of pole searching into the problem of finding roots of Equation 278. We can derive from Equation 277 even more convenient representation of the S-matrix. In order to explain this representation we should notice that the Jost functions J^{+} and J^{-} are functions of the channel wave numbers k_m, i.e., $J^{\pm}(k_1, k_2, \ldots, k_N)$. It is of interest to show what happens when we change the sign of one k_n in J^{-}, i.e., what happens if we put $J^{-}(k_1, k_2, \ldots, -k_n, \ldots, k_N)$. From the definition of the Jost functions (Equation 276) we notice that this change causes the change in the nth channel of the form

$$\varphi^{(l)}_{n,m} = e^{-ik_nr} J^{+}_{n,m}(-k_n) + e^{ik_nr} J^{-}_{n,m}(-k_n) \tag{279}$$

where we have omitted the index l of the Jost functions, which from now on is implicitly assumed. Since $\varphi^{(l)}$ is symmetric to the change $k_n \rightarrow -k_n$, which follows from Equation 241 and the fact that the condition $\varphi^{(l)}(r_0) = 0$ is independent of k_n, the form in Equation 279 is equal to Equation 276. Therefore, $J^{-}_{n,m}(-k_n) = J^{+}_{n,m}(k_n)$, and since other channels are not affected by the change of sign of k_n, we obtain $J^{-}(k_1, k_2, \ldots, -k_n, \ldots, k_N) =$

$J^-(k_1, k_2, \ldots, k_N)$ except that the nth column is replaced by $J^+_{n,m}(k_n)$. From this property of the Jost functions Newton[38,39] showed that the diagonal S-matrix elements are given by

$$S^{(l)}_{m,m} = \frac{D^{(l)}(-k_m)}{D^{(l)}} e^{i\pi(l+1)} \tag{280}$$

where

$$D^{(l)} = \mathrm{Det}[J^-(k_1, k_2, \ldots, k_N)] \tag{281}$$

and the off-diagonals are

$$(S^{(l)}_{m,n})^2 = S^{(l)}_{m,m} S^{(l)}_{n,n} - \frac{D(-k_m, -k_n)}{D} e^{i2\pi(l+1)} \tag{282}$$

In this form the S-matrix elements are directly related to Jost function J^- via the determinant D. From Equations 280 and 282 immediately follows a very important property of the residues β_{l_μ}. The elastic residues are

$$(\beta_{l_\mu})_{m,m} = \frac{D^{(l_\mu)}(-k_m)}{\dfrac{\partial D^{(l_\mu)}}{\partial l}} e^{i\pi(l_\mu+1)} \tag{283}$$

while the inelastic are

$$(\beta_{l_\mu})^2_{m,n} = (\beta_{l_\mu})_{m,m} [(\beta_{l_\mu})_{n,n} \tag{284}$$

therefore, only the diagonal ones are independent.

It should be pointed out that the determinant D contains also the closed channels, while it is implicitly assumed that the matrices J^+ and J^- in Equation 276 refer only to the open channels. This seems to be inconsistent but in the boundary condition of Equation 276 we have assumed that the exponential increase of the closed channel was removed from $\varphi^{(l)}$ by appropriate linear transformation. However, the difference between the value of D when J^- contains the closed channels and when it only contains the open ones, after applying the linear transformation which removes the closed channels, is only a nonessential factor. This factor cancels in Equations 280 and 282. Therefore, the representation of the S-matrix in terms of D has the advantage that open and closed channels are treated as equivalents, which in Equation 277 is not the case.

The wave function $\varphi^{(l)}$ has a particular property if evaluated for $l = l_\mu$. The condition of Equation 278 implies that the equation for X

$$J^- X = 0 \tag{285}$$

has a nontrivial solution, which means that $\varphi^{(l_\mu)} X$ has the asymptotic form

$$\varphi^{(l_\mu)} X = e^{iKr} J^+(l_\mu) X \tag{286}$$

Therefore, there is always a solution of Equation 241 for a Regge pole, which behaves as in Equation 286. This is an important property from which we can prove that Regge poles are always in the region $0 < \arg(l_\mu) < \pi/2$. From the symmetry property of Equation 246 the Regge poles are also found in the region $\tau < \arg(l_\mu) < 3\pi/2$.

Since the Regge pole l_μ is in fact defined as $l_\mu = j_\mu + {}^1/_2$, where j_μ is the root of Equation 278 in the variable l, the set of Equations 241 for the Regge pole l_μ is in the matrix form

$$\varphi'' = \left(V + \frac{l_\mu^2 - 1/4}{r^2} - K^2\right)\varphi + \nu\,\varphi' + \omega\,\varphi \tag{287}$$

Analysis of Regge poles and residues is difficult from Equation 287, but it can be facilitated by transforming the set of Equations 287 into an equivalent form. The transformation is achieved by defining ψ as a transformation of φ, which we write as

$$U\,\psi = \varphi \tag{288}$$

where U is unitary matrix that satisfies the equation

$$U' = \frac{1}{2}\,v\,U \tag{289}$$

with the boundary condition $U \rightarrow I$ for $r \rightarrow \infty$. In such a case ψ satisfies the set of equations

$$\psi'' = \left(W + \frac{l_\mu^2 - 1/4}{r^2} - k^2\right)\psi \tag{290}$$

where

$$W = \tilde{U}(V - K^2)\,U + k^2\,I \tag{291}$$

The two sets of Equations 290 and 287 are equivalent, but Equation 290 is more convenient for analysis. Usually the set of Equations 287 is called the adiabatic and that of 290 the diabatic representation. Their difference is not always a matter of convenience, but also of substantial importance. We will see this in the discussion of perturbation schemes.

From Equation 290 we obtain that $\psi^+ = \tilde{\psi}$ satisfies the set of equations

$$\psi^{+''} = \psi^+\left(W + \frac{l_\mu^{*2} - 1/4}{r^2} - k^2\right) \tag{292}$$

so that when Equation 290 and 292 are combined we find the relationship

$$\psi^+\,\psi'' - \psi^{+''}\,\psi = \frac{l_\mu^2 - l_\mu^{*2}}{r^2}\,\psi^+\,\psi \tag{293}$$

When Equation 293 is integrated between the limits r_0 and infinity, and when Equation 286 is used together with the boundary value of U, then

$$(J^+)^+\,K\,J^+ = 2\,\mathrm{Im}(l_\mu)\,\mathrm{Re}(l_\mu)\int_{r_0}^{\infty}\frac{dr}{r^2}\,\psi^+\,\psi \tag{294}$$

where we have omitted X in Equation 286. Also the plus sign of J^+ should not be confused with the plus sign for the Hermite conjugate. The diagonal values of Equation 294 are

$$\sum_{\text{open}} k_n(J^+)^*_{n,m} J^+_{n,m} = 2 \operatorname{Im}(l_\mu) \operatorname{Re}(l^\mu) \int_{r_0}^{\infty} \frac{dr}{r^2} \sum_n \psi^*_{n,m} \psi_{n,m} \tag{295}$$

Both sums are positive definite from which it follows that either $\operatorname{Re}(l_\mu)$ and $\operatorname{Im}(l_\mu)$ are positive or that they are both negative. This proves that Regge poles can either be in the first or third quadrant of the complex l-plane. Either $\operatorname{Im}(l_\mu)$ or $\operatorname{Re}(l_\mu)$ can be zero, which implies that $J^+_{n,m} = 0$. In such a case we talk about bound states embedded in continuum.[40]

When all channels are closed, then all k_m in Equation 286 are positive imaginary; hence, the left side of Equation 295 is zero. In such a case either $\operatorname{Im}(l_\mu)$ or $\operatorname{Re}(l_\mu)$ is zero; therefore, Regge poles are either real or imaginary.

From all these we conclude that Regge poles for negative energy k^2, i.e., for bound states, are either real or imaginary, and for positive k^2, i.e., for scattering, they are complex and in the first and third quadrant of complex l-plane. More detailed energy dependence of Regge poles can be obtained by taking the derivative of Equation 287 with respect to k^2, which we will designate by a circle above the appropriate function. We obtain

$$\mathring{\varphi}'' = \left[(\mathring{l}^2_\mu)\frac{1}{r^2} - I\right]\varphi + \left(V + \frac{l^2_\mu - 1/4}{r^2} - K^2\right)\mathring{\varphi} + v\,\mathring{\varphi}' + \omega\,\mathring{\varphi} \tag{296}$$

which gives

$$\mathring{\tilde{\varphi}}\,\varphi'' - \mathring{\tilde{\varphi}}''\,\varphi = -\left[(\mathring{l}^2_\mu)\frac{1}{r^2} - I\right]\tilde{\varphi}\,\varphi \tag{297}$$

After integration between limits $r = r_0$ and infinity, and using Equation 286, the energy dependence of l_μ is

$$(\mathring{l}^2_\mu) = \frac{dl^2_\mu}{dk^2} = \frac{\displaystyle\int_{r_0}^{\infty} dr \sum_n \varphi^2_{m,n}}{\displaystyle\int_{r_0}^{\infty} \frac{dr}{r^2} \sum_n \varphi^2_{m,n}} \tag{298}$$

where again for simplicity we have omitted X from Equation 286. The relationship in Equation 298 is valid for any m. More detailed analysis of energy dependence of Regge poles will be done in sections where we will apply the theory. For the moment we have obtained the necessary tool for this analysis.

Let us turn our attention to the properties of residues. From their very definition the S-matrix parametrizes near a pole j_μ as

$$S^{(l)} \sim \frac{\beta_\mu}{l - j_\mu} \tag{299}$$

If the pole j_μ is close to the real axis, i.e., $\operatorname{Im}(j_\mu) \sim 0$, then for real $l = \operatorname{Re}(j_\mu)$ and from the unitarity of the S-matrix we obtain

$$S^{(l)}[S^{(l)}]^+ = \frac{\beta_\mu \beta^+_\mu}{[\operatorname{Im}(j_\mu)]^2} = I \tag{300}$$

or

$$|\beta_{m,m}| \sum_{n} |\beta_{n,n}| = [\mathrm{Im}(j_\mu)]^2 \tag{301}$$

where we have used Equation 284. Since Equation 301 is independent of m we have the summation rule

$$\sum_{m}^{N} |\beta_{m,m}| = \sqrt{N}\, \mathrm{Im}(j_\mu) \tag{302}$$

which means that when $\mathrm{Im}(j_\mu)$ is small the module of $\beta_{m,m}$ is of the order of $\mathrm{Im}(j_\mu)$.

More accurate estimate of the residues is obtained if we take the derivative of Equation 241 with respect to l and calculate

$$\frac{d}{dl}(\tilde{\varphi})\, \varphi'' - \frac{d}{dl}(\tilde{\varphi}'')\, \varphi = -\frac{2l + 1}{r^2}\, \tilde{\varphi}\, \varphi \tag{303}$$

which gives for $l = j_\mu = l_\mu - {}^1/_2$

$$\frac{d}{dl_\mu} J^- = \frac{il_\mu}{2} K^{-1} (\tilde{J}^+)^{-1} \int_{r_0}^{\infty} \frac{dr}{r^2}\, \tilde{\varphi}\, \varphi \tag{304}$$

so that the residue is

$$\beta_\mu = \frac{2}{l_\mu} K^{1/2} J^+ \left(\int_{r_0}^{\infty} \frac{dr}{r^2}\, \tilde{\varphi}\, \varphi \right)^{-1} \tilde{J}^+ K^{1/2} e^{i\pi l_\mu} \tag{305}$$

In general, the residues are complex numbers with the symmetry property $\beta_{-l_\mu} = -\beta_{l_\mu} e^{-2i\pi l_\mu}$. If all channels are closed, then β_μ are either imaginary (for imaginary l_μ) or complex (for real l_μ). for open channels the estimate of β_μ is Equation 302 when $\mathrm{Im}(l_\mu) \sim 0$, but apart from this it is difficult to make any other general estimate of β_μ.

C. Perturbation Theory of Poles and Residues

One of the most powerful techniques in quantum mechanics is the perturbation theory. It is based on the idea that if the Hamiltonian of a system can be separated into two terms, of the form

$$H = H_0 + \epsilon H_1 \tag{306}$$

where H_1 is small and ϵ is a dimensionless parameter of the value between O and 1, then the states of H can be entirely described by the states of H_0. Such an assumption is possible in the quantum theory because it is the theory of linear equations. Since the choice of H_0 depends on many factors there are no general rules regarding how to take H_1. Usually one would assume that the optimal choice of H_1 is that the first correction is small compared to the zeroth order. Even this criterion may not be always correct because what one calls small is not a very well-defined quantity. Sometime it is enough that the correction is a fraction of the unperturbed value (e.g., for the eigenenergies), but sometimes the correction must have a small absolute value, (e.g., for the phase shifts). Very often the perturbation series is not convergent in the ordinary sense, but it is asymptotic.

Our main interest is to apply perturbation theory to the Regge poles and residues. We will first derive a general theory, and later we will decide on a particular choice of parametrization of H. For the moment, however, let us assume that we have split the Hamiltonian into two parts as in Equation 306. In such a case the poles are the roots of equation

$$D(j_\mu;\,\epsilon) = \det[J^-(j_\mu;\,\epsilon)] = 0 \tag{307}$$

where J^-, which was defined in Equation 276, is a function of ϵ. Equation 307 defines an implicit functional relationship between the roots j_μ and ϵ, and we can formally make the expansion

$$j_\mu = j_\mu^{(0)} + \epsilon\, j_\mu^{(1)} + \frac{\epsilon^2}{2!}\, j_\mu^{(2)} + \dots \tag{308}$$

The expansion coefficients are found in the usual way, and the first two are

$$j_\mu^{(1)} = \frac{dj_\mu}{d\,\epsilon};\quad j_\mu^{(2)} = \frac{d^2 j_\mu}{d\,\epsilon^2} \tag{309}$$

where it is assumed that the derivative is calculated for $\epsilon = 0$. From the implicit functional relationship in Equation 307 we obtain[41]

$$j_\mu^{(1)} = -\frac{\dfrac{\partial D}{\partial \epsilon}}{\dfrac{\partial D}{\partial l}} \tag{310}$$

and

$$j_\mu^{(2)} = -\frac{1}{\dfrac{\partial D}{\partial l}}\left\{\frac{\partial^2 D}{\partial \epsilon^2} + 2\frac{\partial^2 D}{\partial \epsilon\, \partial l}\, j_\mu^{(1)} + \frac{\partial^2 D}{\partial l^2}\, [j_\mu^{(1)}]^2\right\} \tag{311}$$

In the same way we can find higher order corrections.

In order to relate the coefficients and the Jost function J^- we will use the equality

$$\det(J) = e^{\mathrm{Tr}[\ln(J)]} \tag{312}$$

from which we obtain

$$j_\mu^{(1)} = -\frac{\mathrm{Tr}\left(J_0^{-1}\dfrac{\partial J}{\partial \epsilon}\right)}{\mathrm{Tr}\left(J_0^{-1}\dfrac{\partial J_0}{\partial l}\right)} \tag{313}$$

and similarly $j_\mu^{(2)}$. We have found formal relationship between the expansion coefficients in Equation 308 and the Jost function J^-. In the next step we must find the relationship between J^-, H_0, and H_1, so that we can relate Equation 309 to perturbation Hamiltonian.

Before doing this, we will find the formal expansion series of the residues since β_μ are also functions of ϵ. Since the matrix elements of β_μ are not independent we only need to consider the diagonal ones, defined by Equation 283. In the expansion of the residues we will not take into consideration the exponential term, since it can be evaluated from Equation 308. Therefore, we will only look for the expansion coefficients in the series

$$\frac{D(-k_m)}{\dfrac{\partial D}{\partial l}} = Q_m^{(0)} + \epsilon Q_m^{(1)} + \frac{\epsilon^2}{2} Q_m^{(2)} + \ldots \tag{314}$$

To find the coefficients, let us assume for the moment that D and $D(-k_m)$ are not evaluated for $l = j_\mu$, but for a value in the vicinity of this pole. In such a case, we can write

$$D = D^{(0)} + (l - j_\mu)\, D^{(1)} + \frac{(l - j_\mu)^2}{2} D^{(2)} + \ldots \tag{315}$$

$D^{(0)}$ is zero, by definition. For $D^{(1)}$ we can write

$$\begin{aligned} D^{(1)} &= D^{(1)}(j_\mu;\ \epsilon) = D^{(1)}[j_\mu^{(0)} + \epsilon j_\mu^{(1)} + \ldots;\ \epsilon] \\ &= D^{(1)}[j_\mu^{(0)};\ 0] + \left[\frac{\partial D^{(1)}}{\partial l} j_\mu^{(1)} + \frac{\partial D^{(1)}}{\partial \epsilon}\right] \epsilon + \ldots \end{aligned} \tag{316}$$

and similarly for $D^{(2)}$. Therefore, the ratio of Equation 314 is

$$\frac{D(-k_m)}{\dfrac{\partial D}{\partial l}} = \frac{D^{(0)}(-k_m) + \left[\dfrac{\partial D^{(0)}(-k_m)}{\partial l} j_\mu^{(1)} + \dfrac{\partial D^{(0)}(-k_m)}{\partial \epsilon}\right] \epsilon + \ldots}{D^{(1)}[j_\mu^{(0)};\ 0] + \left[\dfrac{\partial D^{(1)}}{\partial l} j_\mu^{(1)} + \dfrac{\partial D^{(1)}}{\partial \epsilon}\right] \epsilon + \ldots} \tag{317}$$

from which we obtain the expansion of Equation 314. The leading coefficients are

$$\begin{aligned} Q_m^{(0)} &= \frac{D^0(-k_m)}{D^{(1)}[j_\mu^{(0)};\ 0]}; \\ Q_m^{(1)} &= \frac{1}{D^{(1)}}\left\{\left[\frac{\partial D^{(0)}(-k_m)}{\partial l} - \frac{D^{(0)}(-k_m)}{D^{(1)}} \frac{\partial D^{(1)}}{\partial l}\right] j_\mu^{(1)} + \frac{\partial D^{(0)}(-k_m)}{\partial \epsilon} \right. \\ &\quad \left. - \frac{D^{(0)}(-k_m)}{D^{(1)}} \frac{\partial D^{(1)}}{\partial \epsilon}\right\} \end{aligned} \tag{318}$$

and similarly for $Q^{(2)}$. In the next step we use Equation 312 to relate $Q^{(n)}$ to the Jost function J^- in the same way as it was done in Equation 313. However, the explicit form of this relationship is not essential for our purpose, because in the end we must find how the expansion coefficients $j^{(n)}$ and $Q^{(n)}$ depend on the perturbation Hamiltonian H_1.

We have a choice of two Hamiltonians, and both give the same set of Regge poles and residues. One is adiabatic Hamiltonian, from which we obtain the set of Equations 287, and the other one is diabatic, from which the set of Equations 290 is obtained (l_μ in those sets is replaced by arbitrary l). The diabatic Hamiltonian is most frequently used, because the relevant set of equations, also called the multichannel equations, is simpler. However, this Hamiltonian does not always produce rapidly convergent perturbation series. When this happens one must turn to the adiabatic Hamiltonian.

The multichannel equations for our system are

$$\psi'' = \left[W + \frac{l(l+1)}{r^2} - K^2\right]\psi \tag{319}$$

where we have now defined W so that it goes to zero for $r \to \infty$. The off-diagonal elements of W are taken as perturbation so that Equation 319 can be written as

$$\psi'' = \left[W_0 + \frac{l(l+1)}{r^2} - K^2\right]\psi + \epsilon W^1 \psi \tag{320}$$

By simple inspection, one can show that ψ is given in the form of an integral equation

$$\psi = \psi_0 + \frac{\epsilon}{2i} K^{-1} \int_{r_0}^{r} [f_1(r)\, f_2(r') - f_1(r')\, f_2(r)]\, W^1\, \psi\, dr' \tag{321}$$

where ψ_0, f_1, and f_2 are solutions of Equation 320 for $\epsilon = 0$, but they have different boundary values. ψ_0 is defined as the solution which is zero for $r = r_0$, and the other two functions behave as

$$f_1(r) \sim e^{iKr}$$

$$f_2(r) \sim e^{-iKr} \tag{322}$$

for $r \to \infty$. All three zeroth order solutions are diagonal matrices, because W_0 is diagonal. The function ψ_0 is also called the regular solution and has asymptotic behavior of the form in Equation 276, while f_1 and f_2 are called the irregular solutions.

The solution ψ can be obtained from Equation 321 by iteration, and therefore it is obtained as an expansion

$$\psi = \psi_0 + \epsilon\, \psi_1 + \frac{\epsilon^2}{2}\psi_2 + \ldots \tag{323}$$

For example, ψ_1 is

$$\psi_1 = \frac{1}{2i} K^{-1} \int_{r_0}^{r} [f_1(r)\, f_2(r') - f_1(r')\, f_2(r)]\, W^1\, \psi_0 dr' \tag{324}$$

We will not give the proof here that the series of Equation 323 is absolutely convergent, since this has been shown in various occasions.[14,42] However, it is a comforting thought that we can use Equation 323 as a basis of the perturbation theory without worrying too much about the radius of convergency. When we take the limit $r \to \infty$ of Equation 321 and compare this limit with Equation 276, the Jost function J^- is

$$J^- = J_0^- - \frac{\epsilon}{2i} K^{-1} \int_{r_0}^{\infty} f_1(r)\, W^1\, \psi\, d\, r \tag{325}$$

which is also given as an absolutely convergent expansion in powers of ϵ, when ψ is replaced by Equation 232. J_0^- is an unperturbed Jost function and is a diagonal matrix from which we obtain the unperturbed poles $j_\mu^{(0)}$ as a solution of the equation

$$\det(J_0^-) = (J_0^-)_1\, (J_0^-)_2 \ldots (J_0^-)_N = 0 \tag{326}$$

where $(J_0^-)_n$ is the Jost function which corresponds to the nth channel. Since each of the Jost functions $(J_0^-)_n$ has its own zeros, the poles must carry an additional index which will distinguish the channel where they originate. Therefore the μth pole originating in the nth channel we will write as $j_{\mu,n}$.

We can now calculate the coefficients in Equation 308, for which we need the derivatives of J^- with respect to ϵ and l. They can be easily obtain from Equation 325, e.g., $\partial J/\partial\epsilon$ is

$$J_1^- = \frac{\partial J^-}{\partial \epsilon} = -\frac{1}{2i} K^{-1} \int_{r_0}^{\infty} f_1 W^1 \psi_0 \, d r \tag{327}$$

which has diagonal elements equal to zero. This means that

$$j_{\mu,n}^{(1)} = 0 \tag{328}$$

so that the leading correction of$j_{\mu,n}^{(0)}$ is $j_{\mu,n}^{(2)}$. After taking into account Equation 328, the second order correction Equation 311 is

$$j_{\mu,n}^{(2)} = -\frac{1}{\frac{\partial D}{\partial 1}} \frac{\partial^2 D}{\partial \epsilon^2} = -\left[2k_n(J_0^+)_n \frac{\partial}{\partial 1}(J_0^-)_n\right]^{-1}$$

$$\cdot \sum_{s \neq n}^{N} \frac{2}{k_s(J_0^-)_s} \left(\int_{r_0}^{\infty} \psi\, W_{n,s}^1\, f_{1s}\, dr \int_{r_0}^{r} \psi_{os}\, W_{s,n}^1\, \psi_{on}\, dr'\right) \tag{329}$$

The sum extends over all channels, open and closed, except that the nth channel is omitted. Although the wave function ψ_{0s} is exponentially increasing with r, if s is the closed channel, the coefficient $j_{\mu,m}^{(2)}$ is finite because f_{1s} is asymptotically $\exp(-r|k_s|)$. Higher order coefficients are more complicated, but they are obtained in analogous way.

The perturbation coefficients for the residues are obtained similarly. We distinguish two cases: (1) when the index n of $j_{\mu,n}$ is different than m of $\beta_{m,m}$ and (2) when n = m. In the first case the coefficient $Q_m^{(0)}$ is zero because

$$D^0(-k_m) = (J_0^-)_1 (J_0^-)_2 \cdots (J_0^-)_m \cdots (J_0^+)_n \cdots (J_0^-)_N = 0 \tag{330}$$

since $(J_0^-)_n = 0$. The coefficient $Q_m^{(1)}$ is also zero because of Equation 328 and

$$\frac{\partial D^{(0)}(-k_m)}{\partial \epsilon} = D^0(-k_m)\, \mathrm{Tr}\left\{[J_0^-(-k_m)]^{-1} \frac{\partial}{\partial \epsilon} J^-(-k_m)\right\} = 0 \tag{331}$$

since Equation 327 is zero on diagonal. Therefore, the leading nonzero coefficient is $Q_m^{(2)}$, which is equal to

$$Q_m^{(2)} = \frac{1}{D^{(1)}} \frac{d^2}{d\epsilon^2} D^{(0)}(-k_m) \tag{332}$$

When the second derivative in ϵ is calculated and Equation 328 is taken into account, we obtain

$$Q_m^{(2)} = \frac{1}{D^{(1)}} \left[\frac{\partial^2 D(-k_m)}{\partial \epsilon^2} - \frac{D^{(1)}(-k_m)}{D^{(1)}} \frac{\partial^2 D}{\partial \epsilon^2}\right] \tag{333}$$

where we have used Equation 311. Further derivation gives

$$Q_m^{(2)} = \frac{(J_0^+)_m}{\frac{\partial (J_0^-)_n}{\partial l}(J_0^-)_m} \Big\{ [J_2^-(-k_m)]_{n,m} - (J_2^-)_{n,n}$$
$$- 2 \sum_{s \neq n} [J_1^-(-k_m)]_{n,s} [J_0^-(-k_m)]_s^{-1} [J_1^-(-k_m)]_{s,n}$$
$$- (J_1^-)_{n,s} (J_0^-)_s^{-1} (J_1^-)_{s,n} \tag{334}$$

from where we finally have

$$Q_m^{(2)} = - \frac{1}{4K_nK_m \frac{\partial (J_0^-)_n}{\partial l} (J_0^+)_n (J_0^-)_m^2} \left(\int_{r_0}^{\infty} \psi_n W_{n,m}^1 \psi_{0m} \, dr \right)^2 \tag{335}$$

We obtain a different result when n = m. In such a case $Q_m^{(0)}$ in Equation 314 is nonzero, and it is equal to

$$Q_m^{(0)} = \frac{(J_0^+)_m}{\frac{\partial (J_0^-)_m}{\partial l}} \tag{336}$$

while $Q_m^{(1)}$ is again zero. The second order correction $Q_m^{(2)}$ can be obtained in a way similar to the derivation of Equation 335. Since the resulting expression for $Q_m^{(2)}$ is quite complicated we do not give details of its derivation.

We have more or less completed the most essential parts of the perturbation theory for the poles and residues of the S-matrix, which is based on the diabatic Hamiltonian. The question is how accurate this theory is. It is obvious that for sufficiently small W^1 the theory can adequately describe poles and residues, even with the leading corrections. However, we must also take into account that we are working with systems for which h is small, and therefore we must give more precise meaning to what we call small W^1. For the poles we require that corrections are small in absolute value, because contribution of each pole to the scattering amplitude is in the form cos (πl_μ) (see Equation 264 or 273). Therefore, we require that

$$|j_{\mu,n}^{(2)}| \ll 1 \tag{337}$$

so that the perturbation theory is effective. On the other hand, for the residues it is sufficient that the corrections are small relative to leading term Equation 336. The condition of Equation 337 is more stringent, and therefore the accuracy of this theory is measured by how well Equation 337 is satisfied. The qualitative estimate of Equation 329, in terms of powers of h, is straightforward and gives

$$j_{\mu,n}^{(2)} \sim h^2 \langle W^1 \rangle^2 \sim 0(h^{-1}) \tag{338}$$

where we have made estimate of the integral by the stationary phase method. This means that in the limit of very heavy atom collisions, the diabatic perturbation theory must fail, and therefore it is not adequate for description of atom-atom collisions, except in some isolated cases.

Alternative to diabatic perturbation theory is perturbation theory based on the adiabatic Hamiltonian. An appropriate set is Equations 287, which are

$$\varphi'' = \left[V + \frac{l(l+1)}{r^2} - K^2\right]\varphi + v\varphi' + \omega\varphi \tag{339}$$

where v is antisymmetric and its matrix elements are given by

$$V_{i,j} = \frac{2}{V_i - K_i^2 - (V_j - K_j^2)}\left[U\frac{d}{dr}(W)\tilde{U}\right]_{i,j} \tag{340}$$

while ω is

$$\omega = \frac{1}{2}\frac{d}{dr}v - \frac{1}{4}v^2 \tag{341}$$

Both v and ω are independent of h (see discussion in Chapter 3). Furthermore, typical $v_{i,j}$ is small except in the vicinity of r for which the channels i and j are degenerate, i.e., $W_{i,i} - K_i^2 = W_{j,j} - K_j^2$, where it rapidly changes. We can make qualitative estimate of $v_{i,j}$ in this vicinity. If the off-diagonal elements $W_{i,j}$ are treated as perturbation of the order ϵ then away from the point of degeneracy the matrix elements $v_{i,j}$ are of the order ϵ^0, because $V_i = W_{i,i} + 0(\epsilon^2)$. However, at the point of degeneracy we have an estimate $V_i - K_i^2 \sim V_j - K_j^2 \sim W_{i,i} - K_i^2 + 0(\epsilon)$ from which it follows that $v_{i,j} \sim 0(\epsilon^{-1})$, and therefore in the limit of small ϵ the value of $v_{i,j}$ diverges. On the other hand, the width of the interval, where $v_{i,j}$ is substantial, decreases, and one can estimate that it is of the order ϵ. Overall, we expect that the adiabatic perturbation theory fails in the weak coupling case, i.e., when $W_{i,j}$ are small, but in such a case we can use the diabatic perturbation theory. In that sense, the two perturbation theories are complementary.

If it is assumed that v is perturbation of the order ϵ, then ω can be written as

$$\omega = \frac{\epsilon}{2}v' - \frac{1}{4}\epsilon^2 v^2 \tag{342}$$

The regular solution of Equation 339 can be written as

$$\varphi = \varphi_0 + \frac{1}{2i}K^{-1}\int_{r_0}^{r}[f_1(r)\,f_2(r) - f_1(r')\,f_2(r)] \cdot \left[\epsilon v\varphi' + \frac{\epsilon}{2}v'\varphi - \frac{\epsilon^2}{4}v^2\varphi\right]dr' \tag{343}$$

from where we obtain the Jost function J^-

$$J^- = J_0^- - \frac{1}{2i}K^{-1}\int_{r_0}^{\infty} f_1\left(\epsilon v\varphi' + \frac{\epsilon}{2}v'\varphi - \frac{\epsilon^2}{4}v^2\varphi\right)dr \tag{344}$$

where φ_0 is solution of Equation 339 for $\epsilon = 0$. The zero[th] order poles are obtained from the equation $Det(J_0^-) = 0$, and since v and v' have zero diagonal elements and v^2 is of the order e^2,the first order correction $j_\mu^{(1)}$ is zero, as is the one in diabatic perturbation. The second order correction is obtained in the same way as in the diabatic case in Equation 329, except that care should now be taken because there is an ϵ^2 term in Equation 344. When this is taken into account $j^{(2)}_{\mu,n}$ is given by

$$j^{(2)}_{\mu,n} = -\frac{1}{4i\,k_n \dfrac{\partial}{\partial l}(J_0^-)_n} \int_{r_0}^{\infty} f_{1n}(v^2)_{n,n}\,\varphi_{0n} dr$$
$$- \frac{2}{2k_n (J_0^+)_n \dfrac{\partial}{\partial l}(J_0^-)_n} \left[\int_{r_0}^{\infty} \varphi_0 (v f_1' + \frac{1}{2} v' f_1)\, K^{-1} (J_0^-)^{-1}\, dr \right.$$
$$\left. \cdot \int_{r_0}^{r} \varphi_0 (v\,\varphi_0' + \frac{1}{2} v' \varphi_0)\, dr' \right]^{(n)}_{n,n} \tag{345}$$

where the index (n) of the bracket indicates that the intermediate sum does not involve the channel n.

We can similarly obtain the coefficients in the expansion of residues. The coefficient in Equation 335 is now

$$Q^{(2)}_m = -\frac{1}{4K_n K_m \dfrac{\partial (J_0^-)_n}{\partial l} (J_0^+)_n (J_0^-)^2_m} \left[\int_{r_0}^{\infty} \varphi_{0n} \left(v_{n,m}\,\varphi_{0m}' + \frac{1}{2} v_{n,m}'\,\varphi_{0m} \right) dr \right]^2 \tag{346}$$

The basic difference between the diabatic and adiabatic perturbation theory is that the expansion coefficients of the poles (and also of the residues) behave differently in the limit $h \to 0$. In the diabatic perturbation theory, as we have seen, the nth coefficient increases as h^{-n} while in the adiabatic perturbation theory they are at least finite, which can be directly shown for $j^{(2)}_{\mu,n}$ from Equation 345. Because of this one can also call the adiabatic perturbation theory the semiclassical perturbation theory.[43] However, there are some limitations of this perturbation theory, which were discussed in the chapter on the semiclassical theory and which must be considered if one wants to extend it to inelastic collisions where other than electronic states are excited (e.g., the rotation of molecules).

We did not discuss here a very important case — when the poles are degenerate or nearly degenerate. This is done in Appendix D. The importance of degenerate poles will be demonstrated later.

II. ELASTIC COLLISIONS

A. Properties of Regge Poles and Residues

The simplest collision problem is elastic collision of two atoms. In these collisions the electrons are not excited, and therefore we need only one equation from the set of Equations 237 to describe such processes. Most often they are encountered in low energy collisions of two atoms in their ground electronic state and sometime in collisions of excited atoms, when there is a small probability of changing the electronic state of the system. The relevant equation which we have to solve is

$$\Delta \psi = (V - k^2)\, \psi \tag{347}$$

or the appropriate radial equation

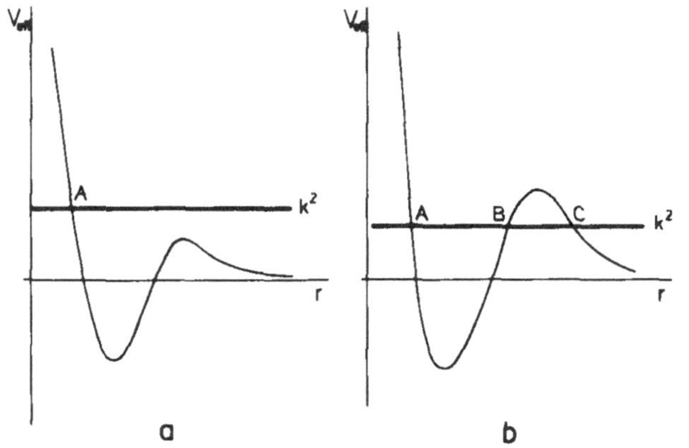

FIGURE 20. (a) For small angular momentum there is only one turning point of the effective potential. (b) Between small and high angular momenta there may be an interval when the effective potential has three turning points.

$$\varphi'' = \left[V + \frac{l(l+1)}{r^2} - k^2\right]\varphi \qquad (348)$$

There is no typical potential V which we could say represents atom-atom interaction. However, most frequently we find potential as shown in Figure 16. Other potentials may be purely, or almost purely, repulsive (e.g., triplet states), or they may have a barrier, as shown in Figure 2 (e.g., $B^1\pi_\mu$ state of Na_2).[44] We will first consider the potentials shown in Figure 16.

In elastic collisions of atoms we can use the WKB or the short wavelength approximation to solve Equation 348. Its basic ideas are described in Appendix B, and we will use here those results to find properties of Regge poles and residues. The central problem in this approximation is finding the points where the relative velocity of two atoms is zero, i.e., we have to find the roots of equation

$$p^2 = k^2 - \frac{(l + 1/2)^2}{r^2} - V = k^2 - V_{eff} \qquad (349)$$

where we have replaced $l(l + 1)$ by $(l + {}^1/_2)^2$, which is the usual procedure in the WKB method. Equation 349 may have either one or three real solutions, which depends on l and k^2. In Figure 20 we show these two cases.

If p^2 has only one turning point one can show that in the simplest WKB approximation the Jost function J^- is

$$J^- = e^{-i\int_A^\infty (p-k)dr + ikA} \qquad (350)$$

which has no zeros for l. On the other hand, when p^2 has three turning points the WKB Jost function is*

$$J^- = -\left[4i\cos\left(\int_A^B p\,dr\right) + \sin\left(\int_A^B p\,dr\right) e^{-2\int_B^C |p|dr}\right] e^{-i\int_C^\infty (p-k)dr + ikC} \qquad (351)$$

* Suitably modified Equation 367 of Reference 6.

which gives approximate roots in l, from the equation

$$\int_A^B p\,dr = \pi(\mu + 1/2); \quad \mu = 0, 1, 2, \ldots \tag{352}$$

if we neglected the exponential factor in Equation 351. The roots j^0_μ of Equation 352 are real, and they can be considered as approximation of the true poles j_μ. We can obtain correction of j^0_μ by using the Newton's root searching method, which gives[28]

$$j^1_\mu = j^0_\mu + \frac{i\, e^{-2\int_B^A |p| dr}}{4j^0_\mu \int_A^B \frac{dr}{r^2\, p}} \tag{353}$$

where again we neglected the exponential factor which appeared in the denominator.

The approximate residues can be obtained from Equation 351 and the fact that for real l the Jost function J^+ is equal to $(J^-)^*$. Hence[28]

$$\beta_\mu = (j^1_\mu - j^0_\mu)\, e^{i\pi(j^0_\mu + 1) + 2i\int_c^\infty (p-k)dr - 2ikc} \tag{354}$$

The estimate in Equation 353 indicates that when p^2 has three turning points the Regge poles are close to the real axis. The imaginary part of these poles is very small, being of the order $\exp(-h^{-1})$, which means that for heavier atoms it goes rapidly to zero. Likewise, the residue of Equation 354 is of the same order of magnitude as $\mathrm{Im}(j^1_\mu)$, which agrees with the finding of Equation 302. Obviously, the estimates of Equations 353 and 354 fail when $B \sim C$, and therefore more accurate calculation, based on the theory which is developed in Appendix B, is needed for obtaining poles and residues.

A more accurate value of the poles is obtained if the WKB approximation is extended to the case when the turning points are complex.[45-47] One way of making such extension is by the use of the theory presented in Appendix B. When we do this properly, then the single turning point formula (Equation 350) for J^- is no longer valid. The Jost function J^- is now a generalization of Equation 351 with the two turning points B and C being complex (for real l they are complex conjugate of each other). In such a form the Jost function has roots, which are continuations of the roots obtained from Equation 353. However, the imaginary part of such poles is the rapidly increasing function of collision energy which means that the poles rapidly go away from the real axis. A typical energy behavior of Regge poles for a potential shown in Figure 16 is given in Figure 21. The curve which shows the energy dependence of a pole is also called a pole trajectory.

Besides these, there are other poles, but they have very large imaginary parts.[28] They are the consequence of the hard core of potential and can be estimated from the Jost function for such potentials. In general, their importance in the scattering amplitude is negligible. This reasoning also applies to entirely repulsive potentials. Under no circumstances do these potentials have three real turning points, and therefore they do not produce poles close to the real axis. They only have poles due to the hard core and in this respect they are not very interesting potentials.

Finally let us consider the potentials which have a barrier, i.e., the one shown in Figure 2. Behavior of Regge poles for such potentials is entirely different from the behavior in Figure 21. We notice that for such potentials we can have three real turning points even when $l = 0$. This happens when k^2 is close to or below the top of the barrier. However, we have seen that Regge poles cannot cross either real or imaginary axis for $k^2 > 0$, and therefore behavior of a typical pole which corresponds to such cases is shown in Figure 22.

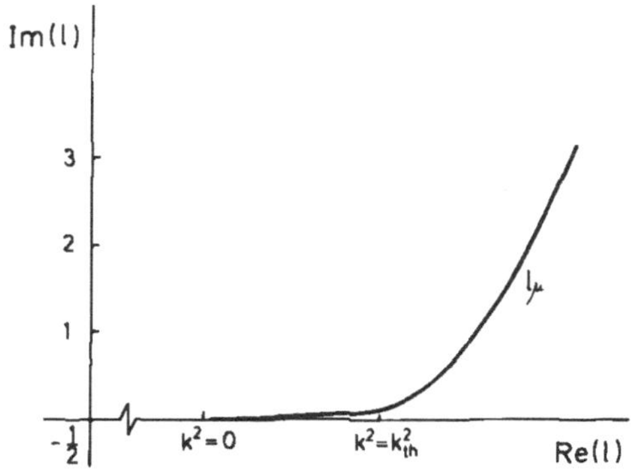

FIGURE 21. Typical energy dependence of Regge pole l_μ for potentials in Figure 16. For $k^2 = k_{th}^2$ two turning points B and C coalesce.

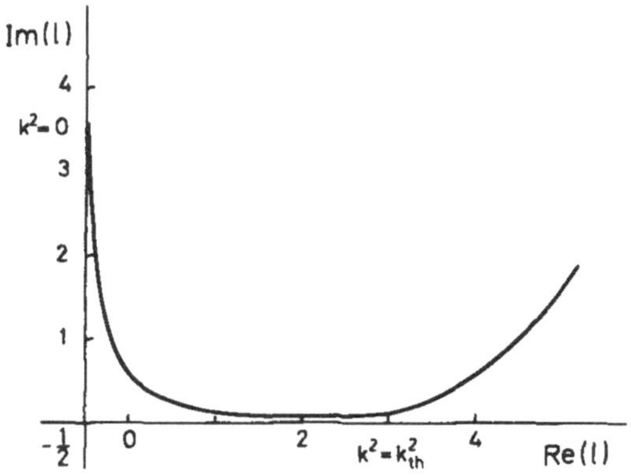

FIGURE 22. Energy dependence of Regge pole for a potential with a barrier.

More accurate k^2-dependence of poles can be obtained from Equation 298, which is for a single channel case

$$\frac{dl_\mu^2}{dk^2} = \frac{1}{\langle r^{-2} \rangle} = C \tag{355}$$

where $<r^{-2}>$ is some average value of r^{-2} for a particular Regge pole. In general, C is a complex quantity, but its feature is that it is a slowly varying function if k^2. Therefore, approximate solution of Equation 355 is

$$l_\mu^2 = C(k^2 - k_0^2) + l_{\mu_0}^2 \tag{356}$$

where we have assumed that for $k^2 = k_0^2$ the Regge pole l_μ has the value $l_{\mu o}$. If $Re(l_{\mu o})$ is large, $[Im(l_{\mu o}]$ is always assumed to be small), which is the case in Figure 21; then l_μ has approximately linear energy dependence

$$l_\mu \sim l_{\mu_0} + \frac{C}{2l_{\mu_0}} (k^2 - k_0^2) \tag{357}$$

On the other hand, when $l_{\mu 0}$ is small, which is the case near the origin in Figure 22, then Equation 356 cannot be reduced to the linear form of Equation 357. In such a case the pole trajectory resembles a hyperbola.

Other relevant information about the poles is their relative spacing. For potentials without natural barrier a typical Regge pole has a large real part obtained from Equation 352. The spacing ϵ between two such neighboring poles is obtained from

$$\int_A^B \left[k^2 - \frac{(l_\mu + \epsilon)^2}{r^2} - V\right]^{1/2} dr - \int_A^B \left(k^2 - \frac{l_\mu^2}{r^2} - V\right)^{1/2} dr = \pi \tag{358}$$

and if it is assumed that ϵ is small then we have approximately

$$\epsilon = -\frac{\pi}{l_\mu} \frac{1}{\displaystyle\int_A^B \frac{dr}{r^2\, p}} \tag{359}$$

This means that spacing between the poles, one of which is shown in Figure 21, is independent of h. On the other hand, for potentials with a natural barrier and if k^2 is below or near the top of the barrier, the real part of Regge pole l_μ is small. In fact Equation 352 gives $l_\mu = 0$ for the μth pole at appropriate collision energy k^2. In such a case the next pole has approximate value $l_{\mu+1} = \epsilon$, which is obtained from Equation 358, if $l_\mu = 0$. However, this time we do not derive Equation 359 but

$$\epsilon^2 = -\frac{2\pi}{\displaystyle\int_A^B \frac{dr}{r^2\, p}} \tag{360}$$

which gives the order of magnitude estimate $\epsilon \sim h^{-1/2}$. We notice that spacing between these poles is now larger than in the previous case, but it only applies for those poles near the origin. Therefore, if there is a pole near origin, then the neighboring ones are far away along the real or imaginary axis.

In general, the Regge poles for potential in Figure 16, and given energy k^2, will be distributed as shown in Figure 23. In the same figure we also show the pole trajectory of each l_μ.

The arrow indicates the pole l_μ for which the function in Equation 349 has still three real turning points when $l = \mathrm{Re}(l_\mu)$. This means that for $l = l_{\mu+1}$ this function has only one real turning point A. Since l_μ are obtained from Equation 352, it follows that there are finite numbers of poles which are close to the real axis. Therefore, above some value of k^2 there will be no Regge poles left near the real axis. The same reasoning applies to the poles for potential in Figure 2. Their typical distribution is shown in Figure 24.

On the real l axis and below the poles in Figure 23, the S-matrix parametrizes as

$$S^{(l)} = \frac{(l - l_0^*)}{(l - l_0)} \frac{(l - l_1^*)}{(l - l_1)} \cdots \frac{(l - l_\mu^*)}{(l - l_\mu)} \cdots e^{i\delta_l} \tag{361}$$

where we have taken into account that $S^{(l)}$ is unitary for real l. The phase δ_l is real and corresponds to the background phase shift, which is approximately obtained from the WKB approximation for a single turning point. If $l < \mathrm{Re}(l_\mu)$, then the turning point is A, and if $l > \mathrm{Re}(l_\mu)$, then it is C. The phase δ_l is a smooth function (the WKB approximation of δ_l has singularity in the vicinity of $l = \mathrm{Re}[l_\mu]$), which has expansion around some $l = L$

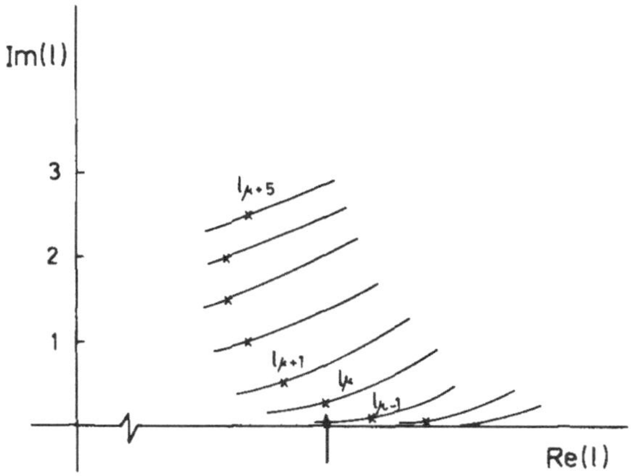

FIGURE 23. Distribution of Regge poles for potential in Figure 16 and at given energy (crosses). The arrow indicates the pole for which C ~ B. The lines show energy dependence of poles.

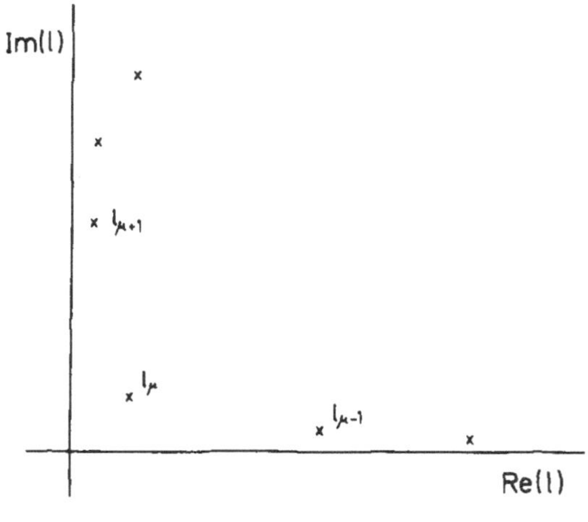

FIGURE 24. Distribution of Regge poles for potential with a barrier.

$$\delta_l = \delta_0 + (l - L)\,\delta_0' + \frac{(l - L)^2}{2}\,\delta_0'' + \frac{(l - L)^3}{6}\,\delta_0''' + \ldots \tag{362}$$

The derivative δ_0' is a so-called deflection function which has the typical shape shown in Figure 25. The solid line represents the WKB approximation, while the dotted line is the exact curve. The value of l_{th} is approximately equal to $Re(l_\mu)$ in Figure 23. Let us assume now that L is equal to $Re(l_{\mu+n})$ and that $l = l_{\mu+n}$. In such a case $Im(\delta_l)$ is

$$Im(\delta_{l_{\mu+n}}) = \delta_0'\,Im(l_{\mu+n}) - \frac{1}{6}\,Im^3(l_{\mu+n})\,\delta_0''' \tag{363}$$

and the estimate of the residue $\beta_{\mu+n}$ is therefore

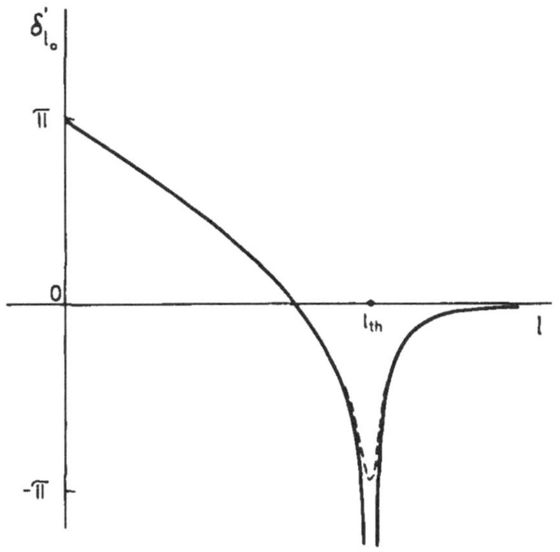

FIGURE 25. Deflection function for potential without a barrier. The WKB approximation has singularity for $l = l_{th}$, while the exact curve (broken line) is finite.

$$\beta_{\mu+n} \sim e^{-\mathrm{Im}(l_{\mu+n})\delta_0' + \frac{1}{6}\mathrm{Im}^2(l_{\mu+n})\delta_0'''} \tag{364}$$

We see from Figure 25 that when n is small the derivative δ_0' is negative and $\delta_0''' > 0$, which means that $\beta_{\mu+n}$ is a rapidly increasing function of n. However, for n moderately large δ_0' is a small negative, while δ_0''' becomes negative, and therefore $\beta_{\mu+n}$ decrease steadily with n. For very large n our analysis becomes less accurate because the approximation in Equation 362 is no longer reliable. The turning point when $\beta_{\mu+n}$ start decreasing is difficult to determine because $\mathrm{Im}(l_{\mu+n})$ is not known, but once the Regge poles are calculated we can obtain this estimate with relative ease.

Similar analysis of the S-matrix and residues is much easier for the potentials in Figure 2. For small l the S-matrix is[48]

$$S^{(l-1/2)} = \frac{l^2 - l_\mu^2}{l^2 - l_{\bar{\mu}}^2} e^{i\delta_{l-1/2}} \tag{365}$$

where l_μ is the pole in Figure 24. In the parametrization in Equation 365 we have also taken into account the symmetry (Equation 246) of the S-matrix. The phase δ_l is obtained in a way similar to Equation 361, but it is also very well approximated by the expansion

$$\delta_l = \pi(l + 1/2) + \delta_0 + \frac{1}{2}(l + 1/2)^2 \delta_0'' \tag{366}$$

where δ_0 and δ_0'' are calculated for $l = -{}^1/_2$. Therefore, we can write for the residue β_μ

$$\beta_\mu = 2i\,\frac{\mathrm{Im}(l_\mu)\,\mathrm{Re}(l_\mu)}{l_\mu}\, e^{i\pi l_\mu + i\delta_0 + \frac{1}{2}\delta_0'' l_\mu^2} \tag{367}$$

and since δ_0'' is small, the residue is also reasonably small.

B. Time-Delay Analysis of Scattering Amplitude

Expansions of the scattering amplitude in Equations 246 and 273 are alternatives to ordinary partial wave representation. We already noticed one advantage of either Equation 246 or 273 over the partial wave expansion. There are fewer Regge poles which significantly contribute to the scattering amplitude than the number of partial waves, while the integrals can be evaluated by methods which are quite accurate and relatively simple. We will see later that these expansions have also physical interpretation which is closer to reality than the partial wave expansion. In that sense Equations 264 and 273 are one way of obtaining Equation 97.

At the moment, however, we will make time-delay analysis of Equations 264 and 273 in order to see what is the essential difference between those two expansions. In both expansions we will assume that $0 \ll \theta \ll \pi$. A typical term in the sum of Equation 264 is

$$f_\mu = \frac{\pi}{k} l_\mu P_{l_\mu - 1/2}(\cos\theta)\, \beta_\mu \frac{e^{i\pi l_\mu}}{\cos(\pi l_\mu)} \tag{368}$$

and for the poles in Figure 23 we can use asymptotic expression for the Legendre function, hence f_μ is proportional to

$$f_\mu \sim \beta_\mu \cos\left(\theta l_\mu - \frac{\pi}{4}\right) \frac{e^{i\pi l_\mu}}{\cos(\pi l_\mu)} \tag{369}$$

The omitted factors in Equation 369 are not essential for time-delay analysis. The partial amplitude in Equation 369 is not yet in the form from which we can obtain the time delay for this component of the incident wave. It is easy to show that f_μ is a rapidly oscillating function of energy, because of the factor which comes from the Legendre function. Therefore, we split f_μ into two components f_μ^+ and f_μ^- which are defined as

$$f_\mu^\pm \sim \beta_\mu \frac{e^{i\pi l_\mu \pm i\theta l_\mu}}{\cos(\pi l_\mu)} \tag{370}$$

so that $f_\mu = f_\mu^+ + f_\mu^-$. The Regge pole l_μ has energy dependence Equation 356, and if we first consider the poles shown in Figure 23, then we can approximately write it as linear dependence (Equation 357). Therefore, as the energy is varied and $\mathrm{Im}(l_\mu)$ is small, the denominator of Equation 370 will be nearly zero whenever $\mathrm{Re}(l_\mu) = n + {}^1/_2$. In the vicinity of such a value of l_μ, the amplitude $f_\mu^\pm$ parametrizes as

$$f_\mu^\pm \sim \beta_\mu\, e^{i\pi l_\mu \pm i\theta l_\mu} \frac{(-1)^n}{\pi\left[i\eta + \dfrac{c}{2n+1}(k^2 - k_0^2)\right]} \tag{371}$$

where we wrote $\eta = \mathrm{Im}(l_{\mu 0})$. This parametrization of $f_\mu^\pm$ resembles the parametrization of the partial amplitude when we studied resonances in Chapter 2, Section II. The factor $(-1)^n$ in Equation 371 comes from the denominator of Equation 370, but if we write $\cos(\pi l_\mu) = \mu^1/_2 e^{-i\pi l_\mu}(1 + e^{2i\pi l_\mu})$ then Equation 370 becomes

$$f_\mu^\pm \sim \beta_\mu \frac{e^{2i\pi l_\mu \pm i\theta l_\mu}}{1 + e^{2i\pi l_\mu}} \tag{372}$$

which near vicinity of every halfinteger value of $\mathrm{Re}(l_\mu)$ has parametrization as in the study of resonances. First, we notice that a single pole describes a series of resonances which we call equivalent resonances, because whenever $Re(l_\mu)$ becomes halfinteger Equation 372 parametrizes as Equation 371. From this it is obvious that parametrization of the scattering amplitude as Equation 264 or 273 is more realistic than the partial wave decomposition because when we consider scattering of wave packets all equivalent resonances are automatically taken into account in the energy integration. As we have seen in Chapter 2 analysis of scattering amplitude within one pole approximation is only qualitatively good because in reality there is usually a large number of resonances. From the partial wave decomposition it is not self-evident that there are, in fact, what we called equivalent resonances. It is intuitively clear from such expansion that if for certain collisions energy resonances appear in some partial waves, then at different energy they will appear in some other partial waves, but it is not clear that resonances at different energy are interconnected or equivalent.

From the theory of resonances in Chapter 2 we obtain that time delay, for the component of the wave packet which is described by $f_\mu^\pm$, is

$$\begin{aligned}\tau_\mu &= \frac{2\mu}{h}\frac{d}{dk^2}[\arg(\beta_\mu) + (2\pi \pm \theta)\,Re(l_\mu)] \\ &= \frac{2\mu}{\hbar}\frac{d}{dk^2}\arg(\beta_\mu) + \frac{2\mu}{\hbar}(2\pi \pm \theta)\frac{C}{Re(l_\mu)}\end{aligned} \tag{373}$$

where we have assumed that $Im(l_\mu) \ll Re(l_\mu)$. From Equation 354 we observe that the term containing β_μ corresponds to the time delay which comes from scattering at the outermost turning point C in Figure 20. In the case when p^2 has only one real turning point for $l = Re(l_\mu)$, then this term corresponds to scattering at A in Figure 20 (in such a case μ_μ is obtained from Equation 361, where δ_l is approximately given by the single turning point WKB formula). In any case, the term with β_μ produces negative time delay, and it is what one calls the direct reflection component of the scattered wave packet.

The coefficient $\mu C/[\hbar\, Re(l_\mu)]$ in Equation 373 can be estimated in the following way. If we replace $\hbar\, Re(l_\mu)$ by $\omega I = \omega\mu <r^2>$, where I is the momentum of inertia of two atoms and notice that C is equal to $<r^{-2}>^{-1} \sim <r^2>$, then this coefficient is ω^{-1}, which means that Equation 373 is

$$\tau_\mu^\pm \sim \tau_d + \frac{2\pi \pm \theta}{\omega} \tag{374}$$

Therefore, time delay for $f_\mu^\pm$ is larger than direct reflection τ_d exactly by the amount which is required by one orbit around the target. By using the classical picture we can say that $f_\mu^\pm$ represent those components of the wave front which follow trajectories as shown in Figure 26.

Let us turn attention to the integral in Equation 264. We will designate this component of f by f_B, and it is

$$f_B = \frac{1}{ik}\int_{i\infty}^{\infty} dl\; l\; P_{l-1/2}(\cos\theta)\; S^{(l-1/2)} \tag{375}$$

The phase of S changes rapidly with l, therefore, it is appropriate to evaluate this integral by the stationary phase method. If again we assume that $0 \ll \theta \ll \pi$, then we can use for $P_{l-1/2}(\cos\theta)$ its asymptotic expansion, in which case f_B splits into two components

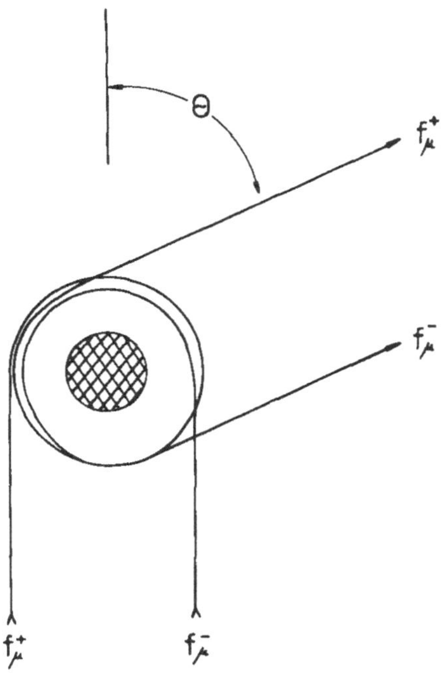

FIGURE 26. Rays of the wave front, which correspond to the scattering amplitudes in Equation 372.

$$f_B^{\pm} \sim \frac{1}{k} \int d\, l \sqrt{l}\, e^{\pm i\theta l + i\delta_{l-1/2}} \tag{376}$$

so that $f_B = f_B^+ + f_B^-$. Let us designate by $l_0^{\pm}$ the value of l for which

$$\pm\theta + \frac{d}{dl}\delta_{l-1/2} = 0 \tag{377}$$

We recognize in Equation 377 the deflection function, which for a potential in Figure 16 is shown in Figure 25. Therefore, Equation 377 has one solution for l_0^- and two solutions for l_0^+, which we designate as l_0^{+1} and l_0^{+2}. In Figure 27 we show trajectories which correspond to the three solutions of Equation 377.

The approximate value of the integral in Equation 376 is proportional to

$$f_B^{\pm} \sim \frac{\sqrt{l_0}}{k\sqrt{\delta''_{l_0}}}\, e^{\pm i\theta l_0 + i\delta_{l_0}} \tag{378}$$

where we have neglected $^1/_2$ compared to l_0. We notice that the value of the integral is independent of h, which means that its dominant contribution is classical. In principle, therefore, the value of f_B could be obtained from classical theory, except the phase. The time delay for these two amplitudes is

$$\tau_B^{\pm} = \frac{2\mu}{\hbar}\frac{d}{dk^2}(\pm\theta l_0^{\pm} + \delta_{l_0^{\pm}})$$

$$= \tau_d^{\pm} \pm \frac{\theta}{\omega^{\pm}} \tag{379}$$

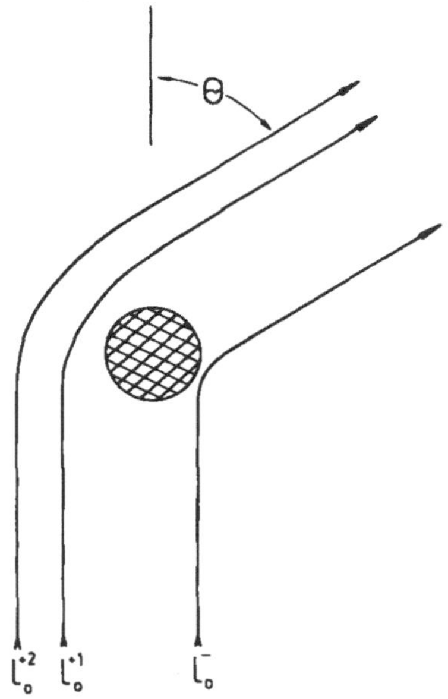

FIGURE 27. Trajectories which correspond to the three solutions of the stationary phase condition in Equation 377.

where $\omega^{\pm}$ is some average angular velocity for a particluar $l_0^{\pm}$. The time delay in Equation 379 is in a form similar to Equation 374, and therefore it has analogous interpretation. The term $\tau_d^{\pm}$ corresponds to the time delay for direct reflection, while the remaining term corresponds to the orbiting of atoms by the angle $\pm\theta$. Figure 27 is the classical illustration of this interpretation.

If we do the same analysis of Equation 273, then we would find that a typical term in the sum of contributions of residues represents a wave delayed by

$$\tau = \tau_d + \frac{\pi \pm (\pi - \theta)}{\omega} \tag{380}$$

In Figure 28a we show trajectories which correspond to the delays in Equation 380.

When the integral along the real axis l in Equation 273 is evaluated by the stationary phase method, then analogous expansion to Equation 377 in this case is

$$-\pi \pm (\pi - \theta) + \frac{d}{dl}\delta_{l-1/2} = 0 \tag{381}$$

which has only solution with + sign. This solution represents the component of the wave which is directly reflected from the core, as shown in Figure 28b. Equation 381 with − sign has no solution because the deflection function has at the most the value of π.

The remaining integral in Equation 273 is negligible.

We can now compare the representations in Equations 264 and 273 in terms of the physical processes which they describe. If we compare Figures 26, 27, and 28, we notice that Equation 273 does not explicitly distinguish contributions of l^{+1} and l^{+2} trajectories in the scattering amplitude. In Equation 264 these two contributions are entirely separated. Also the contribution of f_μ^+ in Figure 26 is not present in Figure 28. It does not mean that because of these differences Representation 273 is less accurate than Representation 264. Between Equations

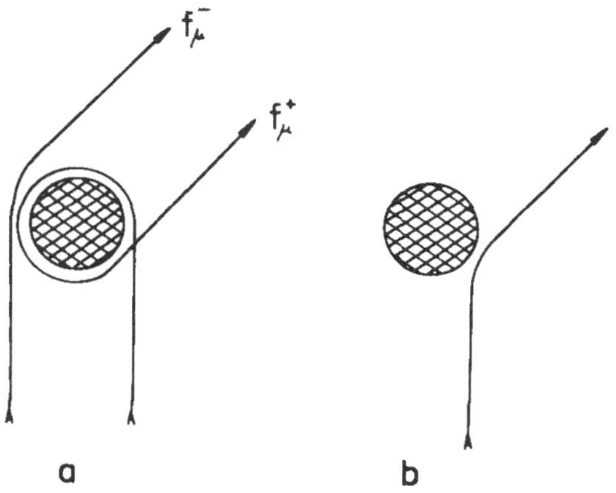

FIGURE 28. (a) Trajectories which correspond to delays (Equation 380). (b) Trajectory for direct reflection, based on Equation 381.

264 and 273 there is an equality sign, and therefore both representations are the exact form of the scattering amplitude. These differences only indicate that contributions of the integral in Equation 264, which are represented by l^{+1} and l^{+2}, are contained in the sum over the residues in Equation 273, but in this sum they are not distinguished separately. In Figure 28a we see that these two contributions are represented by only one trajectory f_μ^-. This also means that the sum over the residues in both Representations 264 and 273 describes also the classical effects (because the contributions of l^{+1} and l^{+2} are classical amplitudes [Equation 378]), and therefore the number of Regge poles increases to infinity in the limit $h \rightarrow \infty$ because each term in this sum is of the order $h^{1/2}$ (if we neglect the estimate of β_μ). In particular, these two sums also describe contributions of higher order orbiting trajectories, i.e., those which go around the target more than once (e.g., three or four times), but these contributions cannot be isolated as separate terms. Therefore, in strict limit $h \rightarrow 0$ Representations 264 and 273 are not adequate because the sums which they contain would be poorly convergent. In such a case one should go back to the Poisson's summation formula and use it before the complex angular momentum analysis was done. Each integral in Equation 245 can be evaluated by the stationary phase method in which case contribution of each orbiting trajectory is separated. However, in such a case one no longer observes the resonance effect, which only confirms that this effect is entirely of quantum origin. Representations 264 and 273 are not at a disadvantage because of their inadequacy in the limit $h \rightarrow 0$. As we have discussed in the chapter on semiclassical theory, the systems are not entirely classical, and therefore the classical theory is at a disadvantage whenever small changes in the initial conditions cause large changes in the outcome of the collision. This is exactly the case for higher order orbiting trajectories. For such trajectories, very small change in the impact parameter (or l) causes very large change in the scattering angle θ, which means that a very narrow piece of the initial wave front would be spread after collision over a wide range of angles (see discussion in chapter on semiclassical theory). Therefore, Representations 264 and 273 bridge the gap between the quantum and classical theory, and because of this they are very useful. In fact, Equation 264 is one further step towards $h \rightarrow 0$ when it is compared to Equation 273. If the stationary phase method, which was used in evaluating Equation 375, fails (because there is no sufficiently wide range of l-s) then one can use Equation 273. However, the contributions of l^{+} and l^{+2} are not then distinguished, which only reflects the quantum behavior.

We have analyzed time delay in elastic collisions when only poles typical of the potential in Figure 16 are present. For this we have made use of Equation 357. However, if we

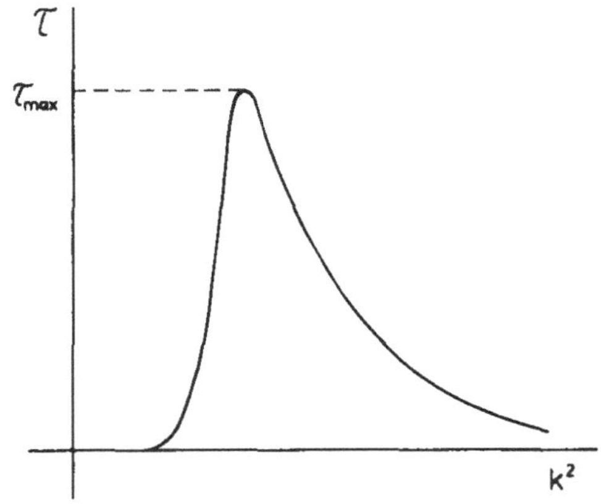

FIGURE 29. Typical time delay for Regge poles with small real part.

analyze time delay when the poles, like the ones in Figure 22, are present, we obtain entirely new results.[49] Let us first consider a typical term from the sum in Equation 264, which is given in Equation 372. For β_μ we can use the approximation in Equation 367 which gives

$$f_\mu^\pm \sim \frac{e^{3i\pi l_\mu \pm i\theta l_\mu + \frac{1}{2}\eta_0'' l_\mu^2 + i\eta_0}}{1 + e^{2i\pi l_\mu}} \tag{382}$$

and the time delay is now

$$\tau_\mu^\pm = \frac{2\mu}{\hbar}\frac{d}{dk^2}\mathrm{Re}\left[(3\pi \pm \theta)\, l_\mu + \eta_0 + \frac{1}{2}\eta_0''\, l_\mu^2\right]$$

$$\sim \frac{2\mu}{\hbar}(3\pi \pm \theta)\frac{d}{dk^2}\mathrm{Re}(l_\mu) + \tau_d \tag{383}$$

where τ_d is negative and corresponds to direct reflection. We cannot use Equation 357 because $l_{\mu o}$ is small, and therefore Equation 383 must be evaluated from Equation 356. If it is assumed that C is real, then Equation 383 is

$$\tau_\mu^\pm \sim \frac{\mu C}{\hbar}(3\pi \pm \theta)\frac{\mathrm{Re}(l_\mu)}{l_\mu l_\mu^*} \tag{384}$$

which has large and positive value, and its typical shape is shown in Figure 29.

If we assume that $\mathrm{Re}(l_{\mu o}) = \mathrm{Im}(l_{\mu o})$, which is not a very serious limitation, then we find that the maximum value of $\tau_\mu^\pm$ in Figure 29 is

$$\tau_{\mu(max)}^\pm \sim \frac{\mu C}{\hbar\, \mathrm{Im}(l_{\mu o})}(3\,\pi \pm \theta) \tag{385}$$

Interpretation of this time delay is that for potentials with a barrier (as shown in Figure 2), the atom, which gets into the scattering region for small impact parameter, orbits around the target with small angular velocity, and therefore it will take a long time to make one and a half orbits, shown in Figure 26. A similar result is obtained if we analyze a typical term in the sum of Equation 273.

It remains to analyze time delay for f_B in Equation 264. Since it is assumed that one Regge pole (e.g., l_μ) is close to the origin, we can write the S-matrix as in Equation 365 so that f_B is

$$f_B \sim \int d\,l\,l\,e^{i\pi l + i\eta} P_{l-1/2}(\cos\theta) + (l_\mu^2 - l_\mu^{*2}) \cdot \int d\,l \frac{l}{l^2 - l_\mu^2} P_{l-1/2}\, e^{i\pi l + i\eta} \tag{386}$$

The first term we recognize as an ordinary background contribution, which comes from the component of the directly reflected wave. We will write it as f_B^0, and the time delay which corresponds to this signal is negative.

The second term we will designate as f_z; it is small but not negligible. Its largest value is when the stationary point of the phase of the integrand coincides with $\mathrm{Re}(l_\mu)$, which is in the backward scattering space because $\mathrm{Re}(l_\mu)$ is small. In such a case we must use Representation 273 instead of Representation 264. The integral in Representation 273 parametrizes in the same way as Equation 386, and if for η we assume the expansion in Equation 366, then f_z is

$$f_z \sim (l_\mu^2 - l_\mu^{*2}) \int_0^\infty \frac{dl\; l}{l^2 - l_\mu^2}\, e^{i\eta_0 + \frac{1}{2}\eta_0'' l^2} P_{l-1/2}(-\cos\theta) \tag{387}$$

There is no simple solution of this integral, but if θ is such that $(\pi - \theta^2/|\eta_0''| \gg 1$ then, as it will be shown in the next section, the amplitude f_z is

$$f_z \sim \exp\left[i\eta_0 + \frac{i}{2}\eta_0''\, l_\mu^2 + i(\pi - \theta)\, l_\mu\right] \tag{388}$$

which is similar to Equation 382 except that the phase $2\pi l_\mu$ is missing. Therefore, f_z describes a delayed wave, i.e., Equation 382, but instead of going once around the target, it only makes the path as shown in Figure 28b. We also notice that the resonance structure is missing in f_z, in contrast to Equation 382.

The meaning of this is that when long-lived states are formed resonances may or may not be formed, because they themselves are not long-lived states, i.e., they are not the cause of positive time delay.

The importance of these results is that one can observe true long-lived states in collisions where the potential has a barrier because the signals which correspond to f_μ and f_z come to the detector with a long delay. In the previous collisions this did not happen because Equation 374 differs only slightly from τ_0. The same is true for Equation 380. In fact, because of such a small time difference in those collisions one is not able to distinguish the arrival of the signals which correspond to f_μ and f_B.

C. Analysis of Cross-Sections

We have seen that with the aid of complex angular momentum analysis the scattering amplitude is conveniently parametrized as a sum in which each term represents a certain type of scattering. From such a parametrization we can directly obtain contribution of these terms in cross-sections. We will first analyze the total cross-section, which is obtained from the optical theorem.

$$\sigma = \frac{4\pi}{k} \mathrm{Im}[f(0)] \tag{389}$$

and when Equation 271 is replaced in Equation 389 we have

$$\sigma = -\frac{8\pi^2}{k^2}\sum_{\mu}^{N} \mathrm{Im}(l_\mu\beta_\mu) - \frac{4\pi^2}{k^2}\sum_{\mu} \mathrm{Im}\left[l_\mu\beta_\mu \frac{e^{i\pi l_\mu}}{\cos(\pi l_\mu)}\right] \tag{390}$$

The first sum is, in fact, approximation of the integral in Equation 271 along the real axis, and its other form is

$$\sigma_B = \frac{2\pi}{k^2}\mathrm{Im}\left[\int_0^\infty dl\ l^2\ S^{(l-1/2)} \frac{d\ \delta_{l-1/2}}{d\ l}\right] \tag{391}$$

which has well-known contribution from the forward glory and diffraction scattering.[50]

A typical contribution of a term in the second sum is

$$\sigma_\mu = -\frac{4\ \pi^2}{k^2}\mathrm{Im}\left[l_\mu\beta_\mu \frac{e^{i\pi l_\mu}}{\cos(\pi l_\mu)}\right] \tag{393}$$

If we use the energy dependence formula (Equation 357) for the poles in Figure 21, and if we assume that $\mathrm{Re}(l_{\mu_0}) = n + {}^1/_2$, where n is an integer, then

$$\sigma_\mu \sim +\frac{4\ \pi^2}{k^2}(n + 1/2)\ \mathrm{Re}\left\{\frac{\beta_\mu e^{i\pi\left[i\eta + \frac{c}{2n+1}(k^2 - k_0^2)\right]}}{\sin\pi\left[i\eta + \frac{c}{2n+1}(k^2 - k_0^2)\right]}\right\} \tag{394}$$

where $\eta = \mathrm{Im}(l_{\mu_0}) \ll 1$. For a small variation of energy this part of the total cross-section is

$$\sigma_\mu \sim \frac{2\pi}{k^2}\frac{(2n + 1)^2}{C}\mathrm{Re}\left[\frac{\beta_\mu}{k^2 - k_0^2 + i\frac{\eta(2n + 1)}{c}}\right] \tag{395}$$

which is the very well-known Breit-Wigner's form of the resonance cross-section.[51,52]

If identical atoms are scattered, then the cross-section in Equation 389 is generalized into

$$\sigma = \frac{4\pi}{k}\mathrm{Im}[f(0) \pm f(\pi)] \tag{396}$$

where the plus sign is for Bose atoms and the minus sign is for Fermi atoms. $f(\pi)$ is easy to obtain from Equation 273 and gives

$$\sigma_\pi = \frac{4\ \pi}{k}\mathrm{Im}[f(\pi)] = \frac{4\ \pi}{k^2}\mathrm{Im}\int_0^\infty d\ l\ l\ S^{(l-1/2)}\ e^{-i\pi l}$$

$$-\frac{4\pi^2}{k^2}\sum_{\mu}\mathrm{Re}\left[\frac{l_\mu\beta_\mu}{\cos(\pi l_\mu)}\right] \tag{397}$$

The integral is the contribution of the backward glory while a typical term in the sum is

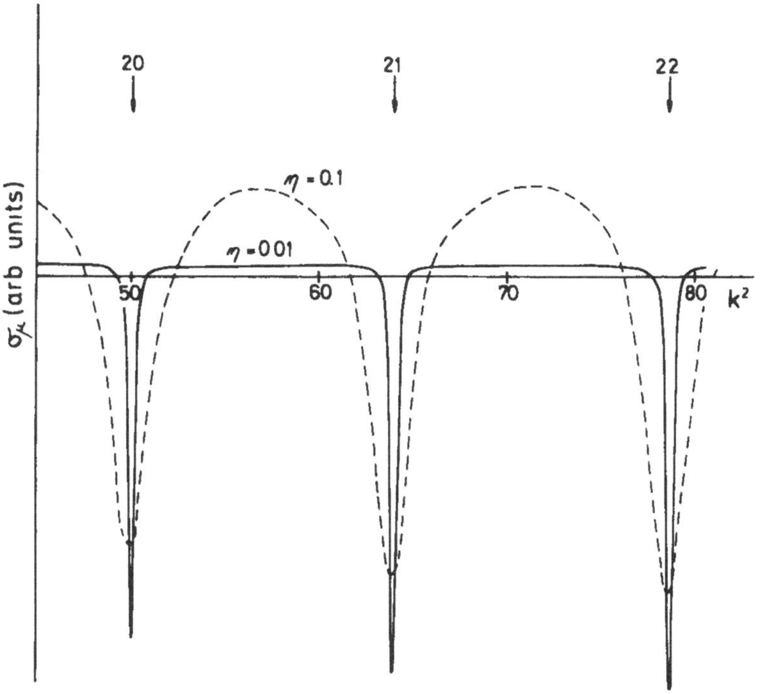

FIGURE 30. Resonance cross-section for pole which is parametrized as in Equation 399. The arrows show position of resonances in a particular partial wave.

$$\sigma_\mu \sim \frac{2\pi}{k^2}\frac{(2n+1)^2}{c}(-1)^n \,\mathrm{Re}\left[\frac{\beta_\mu}{k^2 - k_0^2 + i\eta\dfrac{(2n+1)}{c}}\right] \tag{398}$$

which is equal to Equation 395, except for the factor $(-1)^n$. This only reflects the fact that all odd parity narrow resonances for Bose particles are missing, while for Fermi particles all even parity narrow resonances are missing.

These findings are quite straightforward because for poles in Figure 21 we could use Equation 357. However, when the poles of the type in Figure 22 are considered, then similar analysis is not so easy. In general, we find that the Breit-Winger parametrization (Equation 395) is not valid for these types of resonances.[48,49] We illustrate this by one example in which the Regge pole l_μ is parametrized as

$$l_\mu^2 = (n + 1/2 + i\eta)^2 + 3(k^2 - 50) \tag{399}$$

which is quite realistic, except that in real cases the factor C has small imaginary part. In Figure 30 we show σ_μ for $n = 20$ and two values of η; one is $\eta = 0.01$ (full line), and the other, $\eta = 0.1$ (dotted line). We notice nice resonance behavior.

The same calculation, but with $n = 0$, produces the results shown in Figure 31. It should be pointed out that $\eta = 0.01$ and $\eta = 0.1$ cases give almost identical results; therefore, only one line is shown. The difference between the two figures is immediately noticed. Not only is the spacing between resonances shorter and variable, but resonance for small l do not appear as Breit-Wigner form. Their parametrization is Equation 393, where l_μ is given

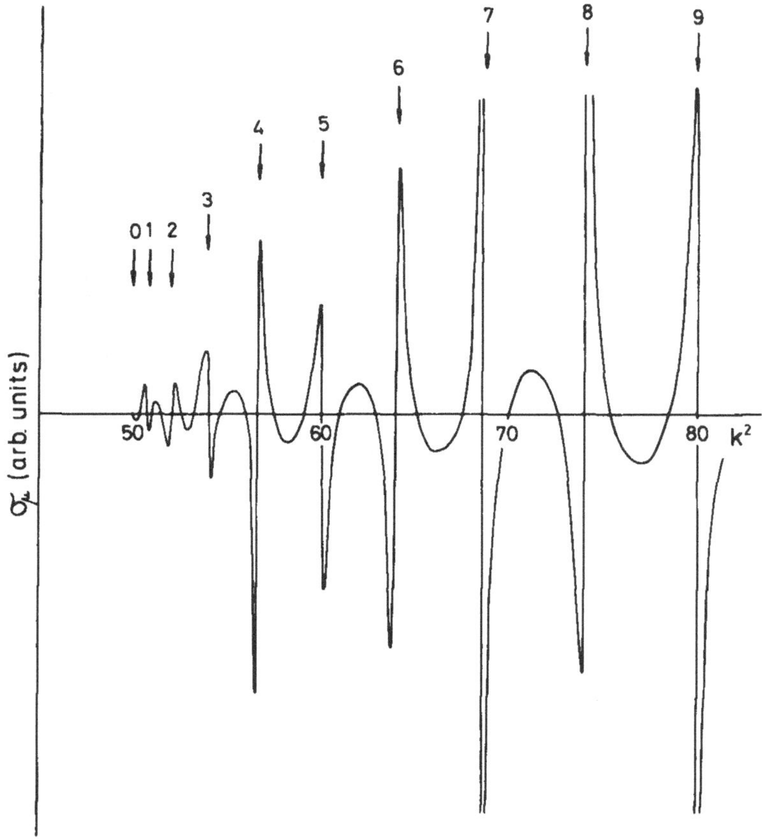

FIGURE 31. Resonance cross-section for pole which has small real part. The pole is parametrized by Equation 399. The arrows show position of resonances in a particular partial wave.

by Equation 356 and cannot be reduced to simpler form. Resonances for larger l have approximate Briet-Wigner form, but their width is narrower than in Figure 30 because n is different in the two cases. This follows from Equation 395, from where the width is $\Gamma = \eta(2n + 1/C \cdot \eta)$. We also notice that the background in Figure 31 is oscillating, while in Figure 30 it is practically constant. Oscillatory background in this case is an indication of the formation of the long-lived state.

The differential cross-section is obtained by taking square module of either Equation 264 or 273. In general, its pattern will not reproduce the pattern of each term in $f(\theta)$, because of additional contributions which correspond to the interference effect between those terms. Because of these interference terms the differential cross-sections have much more complicated patterns than the total cross-sections. Therefore, we will only analyze typical terms in $f(\theta)$. For the Regge poles in Figure 21 the integral in Equation 264 can be evaluated by the stationary phase method because the derivative of the phase of $S^{(l-1/2)}$, with respect to l, is a smoothly varying function of l, as shown in Figure 25. There are three stationary points of the integral, given by solution of Equation 377. Each path in Figure 27 contributes as a separate term in the scattering amplitude, provided that l^{+1} and l^{+2} are well separated, which is usually the case in the forward direction. These contributions have been quite extensively analyzed in literature,[53] and therefore there is no need here to repeat these discussions. We should mention that when $l^{+1} = l^{+2}$ the stationary phase method fails for the integral in Equation 264, and instead one uses uniform approximation.[54] However, this approximation only applies in high energy collisions, i.e., when no Regge poles are close to the real axis, and as result we obtain the well-known rainbow phenomenon in the differential cross-section. In low energy collisions, i.e., when Regge poles are close to the real axis or when the WKB deflection function has singularity for a particular l, one should use the representation Equation 273 for $f(\theta)$. In such a case, as discussed previously, the sum of the residues in Equation 273 describes the rainbow effect in low energy collisions.

Each term in the sum of residues is much easier to analyze. The only θ-dependence is in the Legendre function and for this function we know its approximate behavior. When l_μ is large and $0 \ll \theta \ll \pi$, its asymptotic behavior is

$$P_{l_\mu - 1/2}(\cos\theta) \sim \sqrt{\frac{2}{\pi l_\mu \sin\theta}} \cos\left(l_\mu \theta - \frac{\pi}{4}\right) \tag{400}$$

For small $\mathrm{Im}(l_\mu)$ this function is oscillatory, with very little change in its amplitude, except for the amplitude change because of $\sin^{-1/2}\theta$. On the other hand, when $\mathrm{Im}(l_\mu)$ is large Equation 400 is an exponentially increasing function of θ. Combination of oscillations and exponential increase produces the differential cross-section which corresponds to the f_μ part of the entire scattering amplitude. In general, the result may be quite a complicated interference pattern, as in the case of an H^+-He low energy collision.[56]

Similar discussion applies to Equation 273. The integral has only one point of contribution, and therefore this is what makes Equation 273 more attractive for use than Equation 264. The contribution of residues is similar to that of Equation 264, except that the Legendre function increases towards $\theta \to 0$ for large $\mathrm{Im}(l_\mu)$. Otherwise, the interference pattern is as complicated as in Equation 264.

Another case of interest is the differential cross-section for the case of Regge poles in Figure 22. From now on we will call them zero-angular momentum poles (ZAM poles), because in an interval of collision energy they are in the vicinity of the origin of the l-plane. This case seldom occurs in elastic collisions of atoms, however; it seems that such poles play the essential role in inelastic processes. As we have seen, the dominant contribution of these poles to the scattering amplitude is in the backward scattering space. Therefore, we will make analysis of $f(\theta)$ from Equation 273, and in particular we will evaluate f_B along the real l-axis. Contribution of the other integral is negligible. The amplitude $f_B(\theta)$ parametrizes similarly to Equation 386 for the integral in Equation 264, so that f_B can be written as

$$f_B = f_B^0 + f_Z \tag{401}$$

where f_B^0 is ordinary background contribution without the pole l_μ, and f_Z is

$$f_Z = \frac{l_\mu^2 - l_\mu^{*2}}{k} \int_0^\infty \frac{dl\, l}{l^2 - l_\mu^2} e^{i\eta_0 + \frac{1}{2}\ddot{\eta}_0 l^2} P_{l-1/2}(-\cos\theta) \tag{402}$$

In order to solve this integral we will use the fact that its largest value is near $\theta = \pi$, which also means that small values of l are dominant. This assumption was implicit when we made expansion of the phase shift η in powers of l and when only the first two terms from this expansion were retained. We can also make expansion of the Legendre function in powers of l, which is obtained from its integral representation[55]

$$P_{l-1/2}(\cos \nu) = \frac{\sqrt{2}}{\pi} \int_0^{\nu} \frac{\cos(l\phi)}{(\cos \theta - \cos \nu)^{1/2}} \, d\phi = \frac{\sqrt{2}}{\pi} \sum_{n=0}^{\infty} \frac{(-1)^n}{(2n)!} l^{2n}$$

$$\cdot \int_0^{\nu} \frac{\phi^{2n} d\phi}{(\cos \phi - \cos \nu)^{1/2}} \tag{403}$$

where the angle ν is defined as $\nu = \pi - \theta$ and measures deflection from the backward direction. The integral can be transformed into another form

$$I_n = \nu^{2n+1} \int_0^1 \frac{x^{2n} \, dx}{[\cos(x\,\nu) - \cos \nu]^{1/2}} \tag{404}$$

and if it is assumed that for $n \neq 0$ most of its contribution comes from the vicinity of $x = 1$, then[57]

$$I_n = \frac{\nu^{2n+1}}{(\nu \sin \nu)^{1/2}} \int_0^1 \frac{x^{2n} \, dx}{(1 - x)^{1/2}} = \frac{\nu^{2n+1}}{(\nu \sin \nu)^{1/2}} \frac{[(2n)!]^2 \, 2^{4n+1}}{(4n + 1)!}; \quad n \neq 0 \tag{405}$$

It can be shown by inspection that I_n is approximately[48]

$$I_n = \frac{\pi}{\sqrt{2}} \nu^{2n} P_{-1/2}(\cos \nu) \frac{(2n)!}{2^{2n}(n!)^2} [1 + 0(1/n)] \tag{406}$$

where we have used approximation $P_{-1/2} (\cos \nu) \sim \sqrt{\nu/\sin\nu}$. The other integral is over l, which is of the form

$$I_n = \int_0^{\infty} \frac{dl \; l^{2n+1}}{l^2 - l_{\mu}^2} e^{i/2 \cdot \eta_0' l^2} \tag{407}$$

and it is equal to

$$I_n = 1/2 \int_0^{\infty} \frac{dx \; x^n}{x - x_0} e^{i/2 \eta_0' x} = 1/2 \left(\frac{2i}{\eta_0''}\right)^n n! \, e^{i \frac{\eta_0'}{2} x_0} \int_1^{\infty} dz \, \frac{e^{-\frac{\eta_0'}{2} x_0 z}}{Z^{n+1}} \tag{408}$$

where $x_0 = l_{\mu}^2$. If we now replace both integrals in Equation 402, after making the expansion in Equation 403 we obtain for f_z

$$f_z = \frac{x_0 - x_0^*}{2k} P_{-1/2}(\cos \nu) \, e^{i\eta_0' + i\frac{\eta_0'}{2} x_0} \int_1^{\infty} \frac{dz}{z} e^{-i\frac{\eta_0'}{2} x_0 z - i\frac{\nu^2}{2\eta_0' z}} \tag{409}$$

We can write the integral as $\int^{\infty_1} = \int_0^{\infty} - \int_0^1 = \int_0^{\infty} - \int_1^{\infty}$, where in the last step we have replaced the integration variable z of the second integral by z^{-1}. We obtain finally

$$f_z = \frac{x_0 - x_0^*}{2k} P_{-1/2}(\cos \nu) \, e^{i\eta_0 + i\frac{\eta_0'}{2} x_0} \Big[i\pi \, H_0^{(1)}(\nu l_{\mu})$$

$$- \sum_{n=0}^{\infty} \frac{(-1)^n}{n!} \left(i \frac{\eta_0''}{2} x_0\right)^n E_{n+1}\left(\frac{\nu^2 i}{2\eta_0''}\right) \Big] \tag{410}$$

where we have used the integral representation of the Hankel function $H_0^{(1)}(z)$[57] and the definition of the exponential integral function $E_n(z)$.[55] The sum is rapidly convergent away from $\nu = 0$ because the product $\eta_0'' x_0$ is small and argument of E_{n+1}, large. However, the expression for f_z in Equation 410 is inadequate for $\nu = 0$ and its small vicinity. For $\nu = 0$ we obtain directly from Equation 402

$$f_z = \frac{x_0 - x_0^*}{2k} e^{i\eta_0 + i\frac{\eta_0''}{2}x_0} E_1\left(\frac{i}{2} \eta_0'' x_0\right) \tag{411}$$

Let us examine features of differential cross-section when $Re(l_\mu)$ is such that the product $|\nu l_\mu| > 1$. The amplitude f_z is then approximately

$$f_z \sim e^{i\eta_0 + \frac{1}{2}\eta_0'' l_\mu^2 + i\nu l_\mu} \tag{412}$$

while f_B^0 is

$$f_B^0 \sim e^{il_0\nu + i\eta_0 + i\frac{\eta_0''}{2}l_0^2} = e^{i\eta_0 - i\frac{\nu^2}{2\eta_0''}} \tag{413}$$

where l_0 is determined from the stationary phase condition. An appropriate cross-section is $(f_B)^2$ which contains the interference term $f_B^0 f_z^*$. The phase of the interference term is

$$\arg(f_B^0 f_z^*) \sim -\frac{\nu^2}{2\eta_0''} - \nu\, Re(l_\mu) \tag{414}$$

We notice that the phase has a stationary point for $\nu_s = -\eta_0'' Re(l_\mu) = |\eta_0''| Re(l_\mu)$, which means that around this point oscillations of the differential cross-section are broad. For $\nu \ll \nu_s$ the differential cross-section smoothly varies with ν, while for $\nu \gg \nu_s$ it oscillates very rapidly. A more accurate analysis shows that for $\nu > \nu_s$ the differential cross-section $|F_B|^2$ goes through a broad minimum, and its position is given by

$$\nu_{min} = |\eta_0''| Re(l_\mu) + \left[(\eta_0'')^2 Re^2(l_\mu) + \frac{\pi}{2} |\eta_0''|\right]^{1/2} \tag{415}$$

A typical differential cross-section $|f|^2$ is shown in Figure 32.

We conclude studies of long-lived states in elastic collisions. The only type of long-lived states which we have found are caused by small angular velocity of atoms when their interaction has a barrier. For potentials without a barrier these states are of such a short lifetime that one cannot speak about them as being long lived. As a matter of curiosity it is interesting to notice that if one were able to perform a time-resolved experiment with wave packets, the interference pattern of Figure 32 would not be observed. This pattern is the result of interference of three signals which are separated on the time scale, and it is observed if the incident wave is the plane wave, when these signals cannot be resolved in time.

D. Historical Overview

It is almost certain that the first encounter with long-lived states was made in nuclear physics by Becquerel who observed the decay of the nucleus. The decay must have been quite intriguing because it did not have a simple answer in the classical theory except a phenomenological description.[58] It was observed that an apparently stable nucleus produced

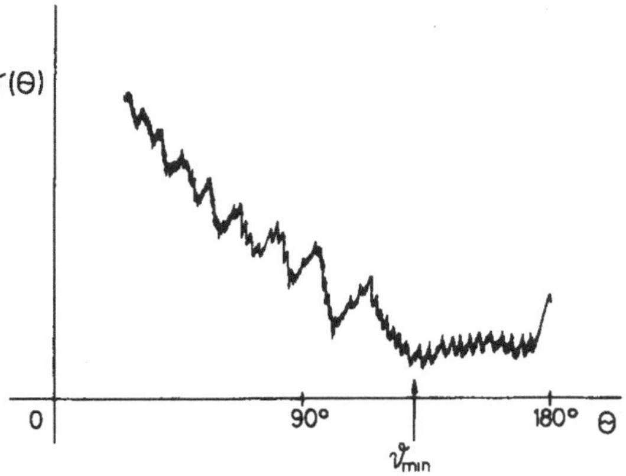

FIGURE 32. Typical differential cross-section for ZAM poles. The arrow shows position of the stationary point (Equation 415).

a daughter nucleus with an extra particle, e.g., an α-particle. The exact moment when the nucleus would decay was not known, but statistics showed that chance of decay fell exponentially with time. The time when this chance fell below a certain value was associated with the lifetime of the nucleus. Explanation of this long-lived state, which is due to Gamow,[59] set the stage for all subsequent events. He explained the decay by introducing the concept of complex energy in the quantum theory. It is arguable whether this was the first use of complex quantities in physics, because they were used earlier in connection with resonance phenomena. However, what Gamow wanted to obtain was solution of the Schrödinger equation for a system of two initially bound particles, which has only the outgoing waves. He thought that only such a solution could represent decay. It is easily shown, as he discovered, that for real energy there is no such solution, but only for complex energy. If formally this complex energy is replaced in the time dependence of the wave function, i.e., for $\psi \sim \exp(-iEt)$, then indeed it is obtained that $|\psi|^2$ decays exponentially with time. Therefore, the imaginary part of this energy was associated with the inverse of the lifetime of the decaying state.

The next development was the study of scattering phenomena in the presence of resonances. It was observed that for certain collision energy the cross-section underwent a rapid change in a small energy interval. Because of this property it was called a resonance effect. Again, the stage was set in nuclear physics where the modeling sequence proceeds in the following way: at collision energies typical of nuclear collisions, the de Broglie wavelength of particles is much larger than the radius of the nucleus (e.g., for 1-MeV proton the de Broglie wave length is $3 \cdot 10^{-4}$ Å which is large compared with 10^{-5} Å, the radius of a typical nucleus); therefore, the cross-section is dominated by one or few partial waves. This means that it is quite sufficient to study resonances in a single partial wave, because only this partial wave affects the cross-section. Indeed, the rapid change of the cross-section was traced to the presence of a complex energy pole of the S-matrix near the real energy axis. From then on when one talks about resonances one usually refers to the complex energy formalism.*[60] Further unification of the theory was achieved when the time-delay concept was introduced into the scattering theory.[9,10] Again the definition of time delay was only useful in the case of a single partial wave, although it was defined also in terms of the scattering amplitude[61] (see Chapter 2 for proper definition). If the time-delay concept is applied to resonances then one obtains that their lifetime is twice their decay time.[62] This result gave formal confirmation to the intuitive belief that decay is one half of time delay. The discovery of energy poles

* In Reference 52, one finds extensive review of early theories of resonance reactions.

of the S-matrix resulted in theoretical description of resonances, e.g., Breit-Wigner parametrization, Argand diagrams, and in fact the entire theory of elementary particles, based on the concept of the S-matrix.

Another, although to a certain degree equivalent, approach to the study of resonances was introduced by Regge.[15] His idea was to transform the partial wave sum by extending the definition of the angular momenta to complex values. A similar idea was used long before Regge, for the case of scattering of electromagnetic waves; however, Regge did not use this method as a step between potential and cross-sections. His intention was to formulate the theory of elementary particles in terms of what are now called Regge poles and avoid using the concept of potential i.e., it was observed, as we have already described, that single pole trajectory describes a series of resonances, and if this trajectory is linear with energy, we have only a two parameter description of a large set of resonances. The idea is very attractive, but primarily as a semiempirical theory of resonances. However, for several reasons Regge's theory was not complete. First, his theory does not give proper definition of total cross-section because his scattering amplitude is only valid away from the forward direction. This may appear a minor point for nuclear processes because only a few partial waves are involved, but for a large number of these waves this inadequacy becomes quite obvious. Furthermore, the background term, i.e., f_B amplitude, is usually neglected, but there is no guarantee that this is always true. For example, for potentials with repulsive core (not necessarily the hard core) this amplitude is far from being negligible. In short, the intention of the theory was not to give another look at resonances, but a convenient way of parametrizing them. Resonances were still treated as complex energy poles of a particular partial wave, and this partial wave determined the features of the cross-sections.

These ideas were adapted to atom and molecular collisions retaining the Regge structure of the scattering amplitude.[34-36] In fact, it can be quite easily shown that his f_B amplitude is infinite for these collisions. Nevertheless, as in the case of nuclear collisions, it was also neglected in atomic collisions, but with little success. Spurious Regge poles had to be introduced in order to remedy the fact that the f_B amplitude was excluded from calculations. The theory had to be reformulated, which was done using the Poisson's summation formula.[28] It must be said though, that this theory is not the theory of resonances, but a prescription how to analyze the scattering amplitude. As we have already seen this theory is quite successful for describing nonresonant phenomena, e.g., f_z amplitude does not have a resonant behavior, but it is crucial to have a Regge pole to parametrize it.

In the end, can we say anything about complex energy vs. complex angular momentum formalism? The representations in Equations 264 and 273 of f are exactly equivalent to the partial wave expansion of f. From these representations it is quite straightforward to obtain contributions of resonances in the cross-sections without, in fact, using the concept of complex energy. On the other hand, in the complex energy formalism one only parametrizes the single partial wave from which it is not clear how this resonance affects the cross-section. For example, a resonance with a moderately large imaginary part of complex energy would, according to the theory, produce the differential cross-section which has the angular dependence of the Legendre polynomial of the partial wave in which it is found. Yet, according to Equation 264 or 273 such a resonance produces oscillatory and exponentially increasing differential cross-section, which is in fact the true behavior.

III. EXCITATION OF ELECTRONS — TWO CHANNEL PROBLEM

The only inelastic process in collisions of two atoms is excitation of electrons. The simplest case is when only two electronic states are involved, e.g., the ground state and some excited state. The dynamics of this collision is described by the two channel Schrödinger equation. If we use the diabatic Hamiltonian, then this equation is in the form

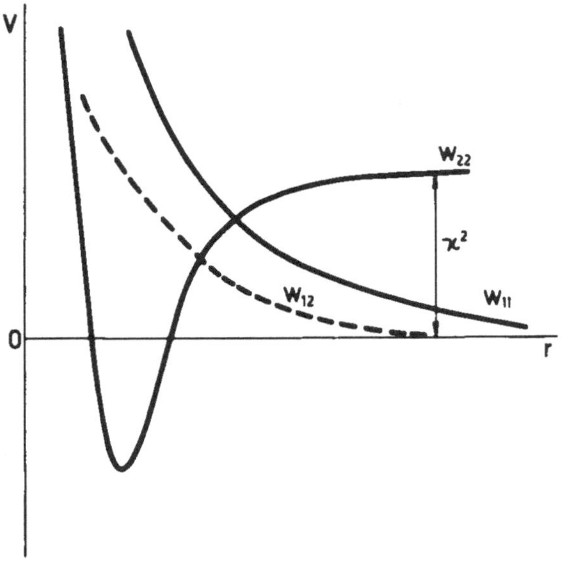

FIGURE 33. Typical set of diabatic potentials for two state problem in electronic excitations.

$$\psi_1'' = \left[W_{11} + \frac{l(l+1)}{r^2} - k^2\right]\psi_1 + W_{12}\,\psi_2$$

$$\psi_2'' = \left[W_{22} + \frac{l(l+1)}{r^2} - k^2\right]\psi + W_{21}\,\psi_1 \qquad (416)$$

where we have separated the angular dependence of the wave function. We will designate the threshold energy of the channel 2 by κ^2. W_{11} and W_{22} can have different shapes, e.g., one of the two shapes discussed earlier in the elastic collision case. However, we will consider the problem as shown in Figure 33, because it contains all the essential elements of the other cases. The coupling term W_{12} is only qualitatively shown and does not necessarily have to have this shape.

We can consider two extreme cases; weak coupling and strong coupling limit. In the first case W_{12} is small, while in the second, W_{12} is large. This distinction is not always very accurate, but it is a first order working hypothesis. More refined distinction will come in actual application of the theory. Let us first discuss the weak coupling case. When W_{12} is neglected, then Equation 416 becomes a set of uncoupled equations, for which the Regge poles can be found from our previous discussion of elastic collisions. Thus, for example, the Regge poles for W_{11} are not close to the real l axis, because this potential is entirely repulsive. This potential has Regge poles with large imaginary part, which are primarily due to the effect of hard core. The contribution of these poles in the scattering amplitude is negligible. On the other hand, Regge poles of the potential W_{22} are quite important, but here we must make distinction between the case when $k^2 > \kappa^2$ and $k^2 < \kappa^2$. When $k^2 > \kappa^2$ the poles correspond to the elastic collision problem for potential in Figure 16 and collision energy $k^2 - \kappa^2$. A typical pole in this case is shown in Figure 21. On the other hand, when $k^2 < \kappa^2$, we must treat the bound state problem for which the Regge poles are either real or imaginary, as discussed earlier in this chapter. A typical set of poles for this case is shown in Figure 34. It should be pointed out that Regge poles are defined as poles of the S-matrix $S^{(l-1/2)}$ and not as the poles of $S^{(l)}$. If the poles are calculated in the variable l then they differ from the Regge poles by one half.

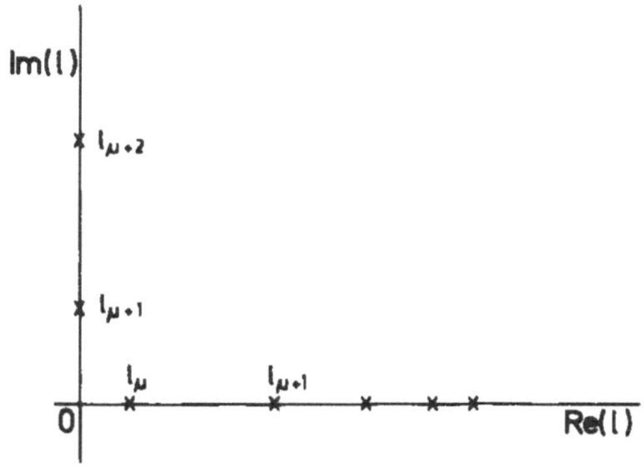

FIGURE 34. Set of Regge poles for bound state problem of potential W_{11}.

The meaning of the Regge pole l_μ in Figure 34 is the following. For given energy $k^2 - \kappa^2$ and $l = 0$, solution in channel 2 does not, in general, coincide with a bound state. However, if l is approximately changed we obtain bound state of energy $k^2 - \kappa^2$ at exactly $l = l_\mu$. Real l_μ means that $k^2 - \kappa^2$ is above this bound state for $l = 0$, while imaginary $l_{\mu+1}$ means that $k^2 - \kappa^2$ is below. In other words, imaginary poles mean that the centrifugal term should be made negative if we want bound states to be lowered to the value $k^2 - \kappa^2$. There is a finite number of real Regge poles, but an infinite number of imaginary poles. This is best seen if we assume that l in Equation 416 is of the form $i\lambda - {}^1/_2$, where λ is large and positive. The potential W_{22} can then be neglected, and the equation for ψ_2 is approximately

$$\psi_2'' = \left(-\frac{\lambda^2}{r^2} - k^2 + \kappa^2\right)\psi_2 \tag{417}$$

The centrifugal term plays now the role of attractive potential which has hard repulsive core at $r = R$. Equation 417 represents a bound state problem, however, in the variable λ instead of k. There is an infinite number of these bound states, which are approximately determined from Equation 352. When this condition is adapted to our case, the poles are solutions of

$$l_\mu \ln \frac{l_\mu^2 + \sqrt{l_\mu^2 - R^2 K^2}}{R K} - \sqrt{l_\mu^2 - R^2 K^2} = \pi n; \quad n = N, N+1, N+2, \ldots \tag{418}$$

where N is a large number and $K^2 = k^2 - \kappa^2$. Usually such poles do not play any role in cross-sections, but near $K \sim 0$ they become significant.

The scattering amplitude for this single open channel case is the same as for elastic collision, described in the previous section, except that now there are two sets of poles; one set contains slightly perturbed poles of the potential W_{11}, while the other contains perturbed poles of potential W_{22}. Perturbation of the poles which belong to W_{11} and W_{22} is directly related to W_{12}, as shown in an earlier section. For small W_{12} it is expected that because of the shape of W_{11} and W_{22}, the poles from the first group will not contribute to the scattering amplitude, while the poles of the second group will be predominantly the ZAM poles. This is because in the vicinity of energy k^2 there is always a bound state of W_{22} for $l = 0$, and hence in the first approximation these poles will be of ZAM character.

We can use previously developed perturbation theory to obtain perturbed poles. If l_μ is a pole, then its expansion in powers of W_{12} is given by

$$l_\mu = l_\mu^{(0)} + \frac{1}{2} l_\mu^{(2)} \tag{419}$$

where $l_\mu^{(0)}$ is a pole from one of the unperturbed channels and $l_\mu^{(2)}$ is given by

$$l_\mu^{(2)} = -\frac{1}{K_m K_s \overset{\circ}{j}{}_m^+ \, j_s^- \, j_m^-} \int_R^\infty \varphi_m W_{ms} f_s^- \, dr \int_R^r \varphi_s W_{sm} \, \varphi_m \, dr' \tag{420}$$

where the index m desginates the channel in which $l_\mu^{(0)}$ is the pole and s is the other channel. For definition of other quantities one should refer to previous sections. As we have said, the poles which belong to channel 1 can be neglected so that the dominant poles will be for m = 2 and s = 1. In particular, we would want to study the ZAM poles, i.e., those for which $l_\mu^{(0)} \sim O$. However, since j_2^- is a function of $(l + {}^1/_2)^2$ (because of hard core of potential) the expansion in Equation 419 fails for these poles because j_2^- is zero and $l_\mu^{(2)}$ is infinite. This problem can be overcome if we define perturbation theory for l_μ^2 instead of for l_μ. We write

$$l_\mu^2 = (l_\mu^2)^{(0)} + \frac{1}{2} (l_\mu^2)^{(2)} \tag{421}$$

and after the use of Equation 419 we find

$$(l_\mu^2)^{(2)} = 2 \, l_\mu^{(0)} \, l_\mu^{(2)} \tag{422}$$

which is finite in the limit $l_\mu^{(0)} \to 0$.

When we now specify m = 2 and $K_2^2 = k^2 - \kappa^2 < 0$ in Equation 420, we obtain for the imaginary part of $(l_\mu^2)^{(2)}$

$$\mathrm{Im}[(l_\mu^2)^{(2)}] = \frac{1}{4k \, |K_2| \, j_2^+ \, \overset{\circ}{j}{}_2^- \, j_1^+ \, j_1^-} \left(\int_R^\infty \varphi_1 W_{12} \, \varphi_2 \, dr \right)^2 \tag{423}$$

where now $\overset{\circ}{j}{}_2^-$ is derivative with respect to $(l + {}^1/_2)^2$. The imaginary part of $(l_\mu^2)^{(2)}$ is always positive because the product $j_2^+ \, \overset{\circ}{j}{}_2^-$ is always positive, and we have assumed that $K_2 = i \, |K_2|$. Real part of $(l_\mu^2)^{(2)}$ does not have such a simple expression and must be calculated directly from Equation 420 and 422. However, in general, we notice that both the real and imaginary parts of $(l_\mu^2)^{(2)}$ depend on the overlap integral between φ_1 and φ_2. Therefore, properties of Regge poles will depend critically on the collision energy. If the collision energy is below the energy where W_{11} and W_{22} cross in Figure 33, the integral in Equation 423 will be small because φ_1 is negligible in the region where φ_2 is oscillatory. On the other hand, when the collision energy is above this point, the integral in Equation 423 is oscillatory and can be estimated by using the WKB solution for φ_1 and φ_2. The approximate value for Equation 423 is then

$$\mathrm{Im}[(l_\mu^2)^{(2)}] \sim \frac{1}{2A_\mu^2} \left[\int_R^\infty \frac{W_{12}}{\sqrt{p_1 \, p_2}} \cos\left(\int_{r_1}^r p_1 \, dr' - \int_{r_2}^r p_2 \, dr' \right) dr \right]^2 \tag{424}$$

In the derivation of Equation 424 we have used (D15) and the WKB approximation of the Jost function j_1^-, given by Equation 436. We have also neglected in Equation 423 rapidly oscillating terms. Most of contribution to the integral in Equation 424 comes from the vicinity of the point $r = r_0$ for which $p_1 = p_2$, i.e., the point where W_{11} and W_{22} cross. The stationary phase method then gives for Equation 424

$$\mathrm{Im}[(l_\mu^2)^{(2)}] \sim \frac{\pi}{2A_\mu^2}\frac{W_{12}^2}{p|W'_{11} - W'_{22}|}\cos^2\left(\delta_0 + \eta\frac{\pi}{4}\right) \tag{425}$$

where $\eta = \mathrm{sign}(W'_{22} - W'_{11})$ and δ_0 is

$$\delta_0 = \int_{r_1}^{r_0} p_1\, dr - \int_{r_2}^{r_0} p_2\, dr \tag{426}$$

In principle Equation 425 can also be zero, in which case l_μ is either real or imaginary, meaning that for this l_μ there is a bound state of the system. Since at this energy one channel is open, i.e., there is access to continuum states, we can talk about a bound state embedded in the continuum. It would appear from this that cross-sections are infinite at this energy, because if for real energy one pole is real and if in addition it has half-integer value, Equation 390 indeed gives infinity. However, as we shall see, the residue at this energy is also zero, and therefore the ratio in Equation 390 will be finite. The meaning of this is that due to the interference effect no part of the incoming wave can penetrate the region of potential W_{22}, and therefore the pole l_μ will remain real.

Perturbation theory gives the residues. Since the unperturbed pole $l_\mu^{(0)}$ is in the closed channel, we obtain that β_μ is

$$\beta_\mu = \frac{\mathrm{Im}[(l_\mu^2)^{(2)}]}{2l_\mu}\frac{j_1^-}{j_1^-}\, e^{i\pi(l_\mu + 1/2)} \tag{427}$$

so that from Equation 390 we can now obtain the appropriate resonance cross-section. General behavior of this cross-section is shown in Figure 31. However, the width of these resonances may be oscillating with energy, which depends on the integral in Equation 423. In the limiting case when $\mathrm{Im}(l_\mu) \to 0$, i.e., for a bound state embedded in the continuum, we can now obtain from Equation 427 that the resonance cross-section goes to zero. In fact, such a state will be also observed when $\mathrm{Re}(l_\mu)$ does not go through half-integer value. The difference is that when $\mathrm{Re}(l_\mu)$ is not half-integer, the resonance cross-section goes through zero without changing sign, i.e., as $(k^2 - k_0^2)^2$, while when it is half-integer, the resonance cross-section changes sign, i.e., it goes as $k^2 - k_0^2$.

The form of residue in Equation 427 greatly resembles Equation 367, and so the time-delay analysis in such collisions gives the same result, which is shown in Figure 29. In fact, we can repeat the same analysis following Equation 382 for the case when the residue is given by Equation 427. Therefore, it is expected that long-lived states will be formed in these collisions, and they are formed because the incident particle is trapped inside the potential W_{22} at nearly zero impact parameter. Appearance of these states in differential cross-section is shown in Figure 32. However, because of the oscillatory nature of $\mathrm{Im}(l_\mu)$ and β_μ (when also bound states embedded in the continuum are possible) the energy dependence of the cross-sections in our case may be different from that in Figures 31 and 32.

We will now consider the opposite extreme case, when coupling W_{12} is large. The equations which we will use for this purpose are based on the adiabatic Hamiltonian and are given by Equation 339. The diagonal matrix V in Equation 339 has elements

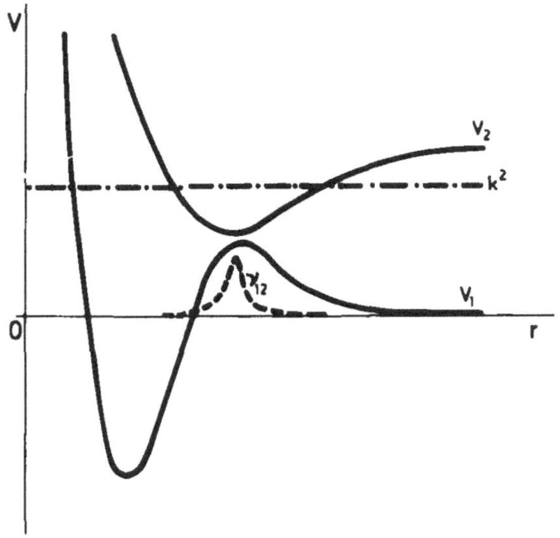

FIGURE 35. Set of adiabatic potentials for the case in Figure 33. The term ν_{12} is coupling in kinetic energy.

$$V_1 = \frac{W_{11} + W_{22}}{2} - \frac{1}{2} [(W_{11} - W_{22})^2 + 4 W_{12}^2]^{1/2}$$

$$V_2 = \frac{W_{11} + W_{22}}{2} + \frac{1}{2} [(W_{11} - W_{22})^2 + 4 W_{12}^2]^{1/2} \tag{428}$$

and are shown in Figure 35. They correspond to the diabatic potentials in Figure 33.

The matrix v, which enters in Equation 339, is antisymmetric, and the only nonzero elements for two channel case are[43]

$$v_{12} = -v_{21} = 2 \frac{(W_{11} - w_{22})^2}{(W_{11} - W_{22})^2 + 4 W_{12}^2} \frac{d}{dr} \left(\frac{W_{11}}{W_{11} - W_{22}} \right) \tag{429}$$

A typical shape of v_{12} is shown in Figure 35 by a broken line. Its maximum is in the vicinity where $W_{11} = W_{22}$ and has the approximate value

$$\max(|v_{12}|) = \left| \frac{W'_{11} - W'_{22}}{2 W_{12}} \right| \tag{430}$$

Perturbation theory with the adiabatic Hamiltonian will be useful if Equation 430 is small. This is obviously the case when $|W_{12}|$ is large, i.e., exactly in the limiting case which we consider. However, we will give shortly a more precise condition when this perturbation expansion is useful.

The zeroth order poles are now determined by V_1 and V_2. All poles of V_2 correspond to bound states (it is assumed that $k^2 < \kappa^2$), and therefore they are real or imaginary. On the other hand, poles of V_1 are complex, and they have the same properties as the poles of potential in Figure 2, which we already discussed. A single pole from this set behaves as the one shown in Figure 22. By looking at Figure 35, we can say which of the poles from these two sets is dominant for description of the scattering amplitude. If k^2 is lower than the bottom of V_2, then all the poles of this potential are imaginary, and therefore their contribution is negligible. On the other hand, when k^2 is above the bottom of V_2, some of these poles are real and must be included in the scattering amplitude. The dominant poles of V_1 are of ZAM character when k^2 is near the top of V; otherwise they are ordinary poles, when k^2 is large.

When perturbation is included then the leading correction to those zeroth order poles is given by Equation 345. The expression is complicated, but it simplifies if we notice that the term involving derivative f_1' is of the order p f_1, where p is local momentum; therefore, it is larger than the rest of the terms which do not involve this derivative. However, such an approximation is not often valid, especially if the turning point of channel 1 coincides with the maximum of v, when f_1 will be relatively small.

If we assume that f_1' is large compared to the other terms, then from Equation 345 we obtain

$$l_\mu^{(2)} = -\frac{1}{2k_2kJ_1J_2^+\ \partial/\partial l\ J_2}\left(\int_R^\infty \varphi_2\, v_{21} f_1'\, dr \int_R^r \varphi_1 v_{12}\, \varphi_2'\, dr' + \int_R^\infty \varphi_2 v_{21}\, \varphi_1'\, dr \int_r^\infty f_1 v_{12}\, \varphi_2'\, dr'\right) \tag{431}$$

where we have assumed that the zeroth order pole is from channel 2. When the pole is from channel 1 then the indexes 1 and 2 replace their role. In order to evaluate the integral in Equation 431 we will use the WKB approximation. This is allowed because the maximum of v_{12} is not near the turning points of V_1 and V_2, by our assumption, and therefore wave functions are quite accurately represented by such a solution. Furthermore, in the vicinity of the point where $W_{11} = W_{22}$, which we designate by r_s, we will approximate W as

$$W = W_0 + (r - r_s)\, W' \tag{432}$$

so that Equation 429 is approximately

$$v_{12} = -\frac{2W_{12}(W_{11}' - W_{22}')}{S_0(r - r_x)^2 + T_0} \tag{433}$$

where

$$S_0 = (W_{11}' - W_{22}')^2 + 4\, W_{12}'^2$$

$$T_0 = 4\, W_{12}^2\left(1 - 4\,\frac{W_{12}'^2}{S_0}\right);\quad r_x = r_s - 4\,\frac{W_{12}\, W_{12}'}{S_0} \tag{434}$$

The WKB approximation gives for f_1

$$f_1 = \sqrt{\frac{k}{p_1}}\exp\left[ikr + i\int_r^\infty (k - p_1)\, dr'\right] \tag{435}$$

and J_1

$$J_1 = \frac{1}{2\sqrt{k}}\exp\left[i\,\frac{\pi}{4} + ikr_1 - i\int_{r_1}^\infty (p_1 - k)\, dr\right] \tag{436}$$

If the WKB solution in Equation 435, together with appropriate solution for φ_1 and φ_2, is replaced in Equation 423, and rapidly oscillating terms neglected, we obtain for the integrals in Equation 431

$$
\begin{aligned}
I_\mu^{(2)} \sim \frac{\sqrt{p}}{8} e^{i\psi_0}\Big[& -e^{2i\Delta_0}\left(\int_R^\infty v_{21} \exp\left(i \int_{r_x}^{r} (p_1 - p_2)\, dr''\right) dr\right)^2 \\
& - \int_R^\infty v_{21}\, dr \int_R^r v_{12} \exp\left(i \int_r^{r'} (p_2 - p_1)\, dr''\right) dr' \\
& - \int_R^\infty v_{21}\, dr \int_r^\infty v_{12} \exp\left(-i \int_r^{r'} (p_2 - p_1)\, dr''\right) dr' \Big]
\end{aligned} \tag{437}
$$

where ψ_0 is the phase of (436), $\mathbf{p}$ is average momentum at r_x and

$$\Delta_0 = \int_{r_1}^{r_x} p_1\, dr - \int_{r_2}^{r_x} p_2\, dr \tag{438}$$

When the integration limits in Equation 437 are changed we obtain

$$
\begin{aligned}
I_\mu^{(2)} = \frac{\sqrt{p}}{8} e^{i\psi_0} \Big\{ & (1 - e^{2i\Delta_0 + i2\delta_0}) \left| \int_R^\infty v_{21} \exp\left[i \int_{r_x}^{r} (p_1 - p_2)\, dr' \right] dr \right|^2 \\
& + 2i \int_R^\infty v_{21}\, dr \int_r^\infty dr'\, v_{12} \sin\left[\int_{r_x}^{r'} (p_2 - p_1)\, dr'' \right] \Big\}
\end{aligned} \tag{439}
$$

where δ_0 is the phase of the first integral in Equation 437. Final expression for Equation 438 is obtained if we use (D15) and Equation 436. In such a case

$$I_\mu^{(2)} = + \frac{i\sqrt{p}}{8\sqrt{k}\, I_\mu^{(0)} A_\mu^2} \left[(1 - e^{2i\Delta_0 + 2i\delta_0})\, M^2 + 2i\, R \right] \tag{440}$$

where M and R are defined by Equation 439. If we use the WKB quantization the phase $2\Delta_0$ can be replaced by

$$2\,\Delta_0 = 2 \int_{r_1}^{r_x} p_1\, dr - \left(\int_{r_2}^{r_x} p_2\, dr - \int_{r_x}^{r_2'} p_2\, dr \right) - \pi(n + 1/2) \tag{441}$$

where r_2' is the turning point of p_2^2. Since p_2^2 is symmetric with respect to r_x, the two integrals in the bracket cancel, and therefore we have approximately for the imaginary part of poles

$$\mathrm{Im}(l_\mu) \sim \frac{\sqrt{p}}{8\sqrt{k}\, I_\mu^{(0)} A_\mu^2} \left[1 - (-1)^n \sin\left(2\delta_0 + 2 \int_{r_1}^{r_x} p_1\, dr \right) \right] M^2 \tag{442}$$

while the shift of the real part is

$$\Delta\,\mathrm{Re}(l_\mu) = -\frac{\sqrt{p}}{8\sqrt{k}\, I_\mu^{(0)} A_\mu^2} \left[2R + (-1)^n M^2 \cos\left(2\delta_0 + 2 \int_{r_1}^{r_x} p_1 dr \right) \right] \tag{443}$$

We will now estimate the integrals M and R from the linear approximations in Equations 432 and 433. It can be shown that

$$\int_{r_x}^{r} (p_1 - p_2)\, dr' \sim \frac{1}{p} \int_0^{r-r_x} \sqrt{T_0 + S_0 y^2}\, dy \tag{444}$$

so that the integral M is approximately

$$M = \int_{-\infty}^{\infty} \frac{dx}{1 + x^2} \exp\left(i \frac{T_0}{p\sqrt{S_0}} \int_0^x \sqrt{1 + y^2}\, dy \right) \sim \frac{2\pi}{3} e^{-\pi T_0/4p\sqrt{S_0}} \tag{445}$$

Estimate of the integral in Equation 445 is only an order of magnitude accurate, but it shows the trend of exponential decay of M with h^{-1}. One establishes easily that $\delta_0 = 0$ in Equation 440. Estimate of the double integral R gives that it is smaller than M^2 so that in the first instance we can neglect it. Finally, we obtain for the correction $(l_\mu^2)^{(2)}$, which is defined by Equation 421,

$$(l_\mu^2)^{(2)} \sim \frac{i\sqrt{p}}{\sqrt{k}\, A_\mu^2} e^{-\pi T_0/2P\sqrt{S_0}} (1 - e^{2i\Delta_0}) \tag{446}$$

Both coefficients in Equations 425 and 446 are corrections to the unperturbed poles, but as we have assumed, one is valid for small W_{12} and the other for large W_{12}. By comparing these two expressions we obtain more accurate definition of the weak and strong coupling limits. Weak coupling, or diabatic limit, is valid when

$$\frac{W_{12}^2}{p\,|W'_{11} - W'_{22}|} \ll 1 \tag{447}$$

while in the opposite case we talk about the strong coupling, or adiabatic limit. We notice that the condition in Equation 447 is equivalent to $h^{-1} \ll 1$, which is never satisfied in the semiclassical limit. Therefore, it is expected that diabatic perturbation theory for electronic transition is of limited use in the collisions of heavy atoms.

The residue is given by Equation 427, and if we use Equation 446, we obtain

$$\beta_\mu \sim \frac{\sqrt{p}\, i}{2\sqrt{k}\, A_\mu^2\, l_\mu^{(0)}} e^{-\pi T_0/2p\sqrt{S_0}} \frac{j_1^+}{j_1^-} e^{i\pi l_\mu}[1 - \cos(2\Delta_0)] \tag{448}$$

With this we have obtained all the necessary data to obtain the cross-sections which result from such behavior of poles. We will analyze the case when $\mathrm{Re}(l_\mu^{(0)})$ is sufficiently large so that the linear approximation in Equation 419 is applicable. Therefore, the pole l_μ is given by

$$l_\mu = l_\mu^{(0)} + i \frac{C}{l_\mu^{(0)}} (1 - e^{2i\Delta_0}) \tag{449}$$

where C is a factor which is not strongly energy dependent. The phase Δ_0 is a slowly varying function of energy, which we can show by analyzing its energy derivative. From Equation 448 we have

$$\frac{d}{dk^2}(2\Delta_0) \sim 2\frac{d}{dk^2} \int_{r_1}^{r_x} p_1\, dr = \int_{r_1}^{r_x} \frac{dr}{p_1} - \frac{dl_\mu^{(0)2}}{dk^2} \int_{r_1}^{r_x} \frac{dr}{r^2 p_1} \tag{450}$$

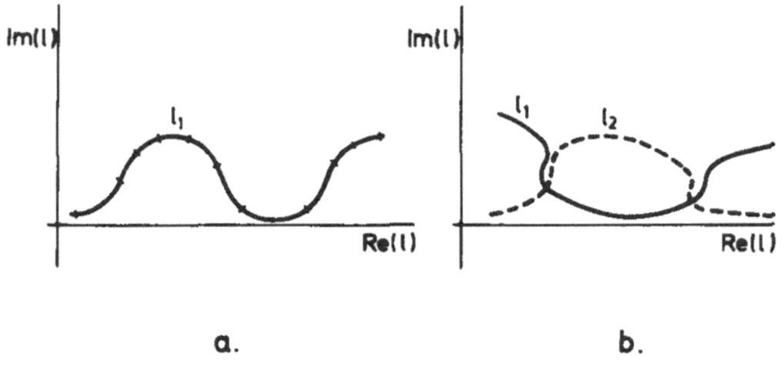

FIGURE 36. (a) Typical energy dependence of single pole in two state problem. Equal energy intervals are shown by marks. (b) Realistic energy behavior of poles in two state problem.

which indeed gives small value for the derivative of Δ_0 if we use the relationship in Equation 298 and the WKB solution for φ_2.

Equation 449 is the equation for the cycloide type curve, which means that we can imagine the pole l_μ being on the perimeter of a circle which rolls on the real l axis, with "speed" $dl_\mu^{(0)}/dk^2$ and radius $C/l_{\mu(0)}$. The radius of the circle is variable, i.e., it is decreasing with energy. However, the angular velocity of the "rolling circle" is not always in the direction which its "translational velocity" suggests, i.e., it is not always negative. The sign of the "angular velocity" is determined from Equation 450. Usually if r_1 is larger than r_2 then Equation 450 is positive, and for this case we show typical behavior of l_μ in Figure 36a. Separation of the marks on the curve designates equal energy intervals.

Such behavior of a pole is altered when other poles are taken into account. We notice from Equation 443 that the shift in Equation 449 is not the same for all poles. Two neighboring poles can, in fact, be shifted towards each other so that their paths cross for a particular energy. This means that at this point the S-matrix does not have simple pole, but the second order singularity. However, second order poles are rather unrealistic for the S-matrices, but there is no general theorem which excludes their existence. Nevertheless, we will reject the idea that our perturbation result is correct, and instead we will assume that in the vicinity of this collision energy the two curves do not cross, but they behave as avoided crossing in degenerate perturbation theory. We should emphasize though that this situation is different from the degenerate case because two poles cross in the same and not in two different channels. Proper perturbation theory for such poles is more complicated than ordinary theory, and therefore we only mention that one should consider using it when ϵ between channels is sufficiently large so that the second order expansion in Equation 419 produces degenerate poles. Realistic behavior of two poles is shown in Figure 36b.

The cross-section which results from such behavior of poles is quite interesting. Since energy derivative of $l_\mu^{(0)}$ is relatively large, succession of resonances is very rapid, very similar to the case of ZAM resonances. Therefore, the overall appearance of total cross-sections will be as shown in Figure 31. However, we must also consider the oscillatory, imaginary part of l_μ so that when it is large the cross-section σ_μ will be small. Also, when $\text{Im}(l_\mu)$ is very small the cross-section σ_μ is negligible. This means that only in the energy interval when $\text{Im}(l_\mu)$ has intermediate values will the cross-section σ_μ have nonnegligible value. The result of this is shown in Figure 37. The cross-section σ_μ appears as a series of pulses, and each contains several resonances with rapidly oscillatory background, which is an indication of formation of long-lived states. As we have discussed earlier, in the case of ZAM resonances, because of such background individual resonances are not easily resolved.

Time delay for such collisions is essentially obtained from the derivative of $\text{Re}(l_\mu)$, because all other terms in Equation 448 are slowly varying functions of k^2 (except, of course, the

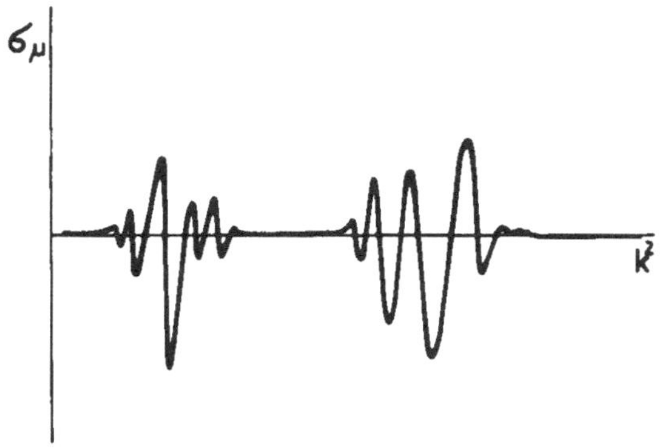

FIGURE 37. Energy dependence of resonance total cross section for a pole in Figure 36b. The cross-section appears as a series of pulses, rather than a uniform sequence of resonances.

phase of j_l^+, which is included in the time-delay analysis anyway). Therefore, we obtain similar result as in Equation 384 except that the time delay is slowly varying with k^2 because of functional dependence (Equation 443). However, near the point where the two poles meet, the time delay changes rapidly and can even be very negative. In such a case the cross-section σ_μ becomes negligible, thus in no way the causality principle is violated.

In this section we have discussed only one possible case involving two electronic states in collisions. This particular example was chosen because it contains all the essential elements of a more general case, except the energy transfer.

Chapter 6

ATOM-MOLECULE SYSTEMS

I. SCATTERING ON SPHERICAL POTENTIAL

A. Vibrationally Inelastic Collisions

The simplest collision process of atom and molecule is when no exchange of atoms (nonreactive collision) and no electronic states of either atom or molecule are involved. The only energy exchange in such collisions is through the rotations and vibrations of molecules, which very often happen simultaneously. Therefore, strictly speaking, vibrations and rotations cannot be considered independently, but in many cases the vibrational quantum of the molecules is so large that in a certain energy range it is sufficient to treat only the rotational energy transfer. For example, the first vibrational state of CO is about 0.27 eV above its ground state, so that in any collision in which this energy is not available the only energy exchange will be through rotations of molecules. Of course, we would have to take into account the distortion of the molecule in its high rotational state, but with great accuracy we can assume that the molecule is rigid below the first vibrational threshold. Very often, vibrational excitation of molecules is a very inefficient process (e.g., if incoming atoms are very light and the atoms of the molecule are heavy) so that vibrations can be neglected even when the collision energy is above the first vibrational threshold. Usually in such a case the rotational energy change is also a very inefficient process. Nevertheless, we will first consider the vibrational energy transfer in atom-molecule collisions, because this process is theoretically easier to formulate than rotational energy transfer. Of course, in the modeling of vibrational energy transfer we will neglect the contribution of rotational degrees of freedom. This approximation is highly speculative, but it is believed that it reasonably well describes certain collision processes.

In collisions of atoms and molecules there are essentially two ways in which the vibrational energy of the molecule can be changed: one which we will call the kinematic mode and the other, the dynamic mode. The first mode relies on the fact that when the atom hits the molecule the energy exchange occurs because of the direct impact of this atom with one of the atoms of the molecule. This mode is very similar to the energy exchange between billiard balls and depends entirely on the ratio of masses between the two impacting atoms. The dynamic mode is entirely due to the form of interaction between atom and molecule, prior to the short repulsive region. For example, if the impacting atom is an ion it will polarize the molecule so that this molecule will be distorted from its original geometry. When the atom leaves the interaction region it will leave the molecule in a different vibrational state, meaning also that the kinetic energy of the atom will differ from its original value. The two energy transfer modes always go together, but under some circumstances one dominates over the other, e.g., if the whole system interacts entirely through the Van der Waals forces, the dominant mode will be kinematic. On the other hand, if the incoming atom is a very light ion, then the mode will be predominantly dynamic.

Very often for atom-diatom collisions one uses the collinear model because, it is argued, most of vibrational energy change occurs in such a configuration, at least when the kinematic mode is dominant. Also, because the theoretical description of this model is relatively simple, it is very popular for describing vibrational processes. In addition, this model is also used for testing new theoretical techniques, and therefore it is unavoidable when atom-molecule collisions are considered. However, with all its merits one should not neglect some of the very serious drawbacks of the collinear model, which make it nothing more than an interesting exercise and sometimes an order of magnitude estimate of vibrational energy transfer. In particular, if we want to study long-lived states and resonances in such collisions the results may be even misleading. The collinear model is equivalent to single partial wave scattering (equivalent of an S-wave scattering), and therefore resonances which are observed have appearances similar to resonances of certain partial waves in full three-dimensional treatment. However, we have seen in the case of zero-angular momentum (ZAM) resonances in Chapter 5, Section II.C that their appearance in cross-section is entirely different than what would be predicted from analysis of the same resonances in partial waves. Therefore, we must use an adequate alternative option which is closer to reality for describing long-lived states than the collinear model.

The assumption that in collision the rotation of molecules is not changed implies that we can neglect the total rotational angular momentum operator in the Hamiltonian of the molecule. We can assume that molecule is not rotating and that its orientation is fixed in space. The Schrödinger equation which describes such a collision is

$$(T_\rho + V_M)\,\psi + [T_A + V(\mathbf{r}, \rho)]\,\psi = E\,\psi \tag{451}$$

where T_ρ is the kinetic energy of atoms in a molecule, from which the rotation of the molecule, as a whole, is neglected. V_M is the potential for atoms in an isolated molecule, T_A is the kinetic energy of the incident atom, and $V(\mathbf{r}, \rho)$ is the potential between the atom and molecule. If by φ_n we designate the eigenfunctions of the Hamiltonian of the molecule, then Equation 451 is equivalent to a set of equations

$$T_A\,\omega_n + \Sigma\,\langle\varphi_n|V|\varphi_m\rangle\,\omega_m = (E - E_n)\,\omega_n \tag{452}$$

where the matrix elements of $V(r, \rho)$ are not spherically symmetric because they depend parametrically on the orientation of the molecule. In order to be more specific we will consider scattering on diatomic molecule in which case ρ plays the role of interatomic separation in the molecule. The potential V is a function of r, ρ and the angle α, which is the angle of relative orientation of the molecule with respect to the incoming atom. This angle is given by $\cos\alpha = \cos\theta\cos\theta_M + \sin\theta\sin\theta_M\cos(\phi - \phi_M)$ where θ and ϕ are polar angles of the axis of molecule. If we replace ω_n in Equation 452 by the expression

$$\omega_n = \frac{1}{r}\sum_{l,m}\psi_{l,m}^{(n)}(r)\,Y_{l,m}(\theta, \phi) \tag{453}$$

where $Y_{l,m}(\theta, \phi)$ are spherical harmonics, then the set of equations for $\psi_{l,m}^{(n)}(r)$ is

$$-\frac{\hbar^2}{2\mu}\left[\frac{d^2}{dr^2}\psi_{l,m}^{(n)} - \frac{l(l+1)}{r^2}\psi_{l,m}^{(n)}\right] + \sum_{l',m',n'} V_{l,m,l',m'}^{(n,n')}(r;\ \theta_M, \phi_M)\,\psi_{l',m'}^{(n')} = (E - E_n)\,\psi_{l,m}^{(n)} \tag{454}$$

where μ is the reduced mass of the system. When $\theta_M = 0$ or $\theta_M = \pi$, the set of Equations 454 is a three-dimensional version of the collinear model. We notice that now all l are coupled, and solving this set requires special attention. This is because in the usual case, when l are not coupled, the initial value of $d/dr\,\psi_{l,m}^{(n)}$ at $r = r_0$ can have arbitrary value since when $\psi_{l,m}^{(n)}$ are matched with the scattering boundary condition, the proper normalization constant of $\psi_{l,m}^{(n)}$ is automatically determined. In the case of Equation 454 we cannot assign arbitrary value to this derivative because the solution $\psi_{l,m}^{(n)}$ does not have simple dependence on the choice of this initial condition. The way to solve this problem we will demonstrate on a simpler problem of elastic collisions. In such a case Equation 454 becomes

$$\psi_l'' = \left[V + \frac{l(l+1)}{r^2} - k^2\right]\psi_l + \sum_{l'=0}^{L} v_{l,l'}\,\psi_{l'} \tag{455}$$

We will define a set of L + 1 linearly independent solutions, which differ one from the other in the choice for $\psi'_l(r_0)$ (for all of them $\psi_l[r_0] = 0$). Additional index l_0 of ψ_l designates the solution for which all $\psi'_l(r_0)$ are zero except $\psi'_{l_0}(r_0)$. Therefore, when we match the scattering condition and the solutions of Equation 455 we will have

$$e^{ikz} + \frac{1}{r}e^{ikr}f(\theta) = \sum_{l_0=0}^{L} a_{l_0}\cdot\sum_{l=0}^{L}\psi_l^{(l_0)}P_l(\cos\theta) \tag{456}$$

where a_{l_0} are constants which need to be determined. When we match Equation 456 in the usual way, we obtain for the S-matrix in the lth partial wave

$$S_l = \frac{1}{2l+1}\sum_{l',l_0=0}^{L} j^{+}_{l,l_0}(j^{-})^{-1}_{l_0,l'}(2l'+1)(-1)^{l'+1} \tag{457}$$

which replaces S_l in Equation 243, of course for a single channel case. Similarly, we proceed in the case of Equation 454. However, we already see in Equation 457 certain difficulties if we want to implement the ideas of complex angular momentum theory to nonspherical potentials. The problem is how to make l complex in Equation 457. We will postpone for the moment this discussion and instead discuss a simpler model in which there is no coupling between l, so that the coupling matrix elements in Equation 454 are independent of l and θ_M. This model is also known as the breathing sphere or vibrating sphere model,[63] and the set of equations is

$$\psi_n'' = \left[v_{n,n} + \frac{l(l+1)}{r^2} - k_n^2\right]\psi_n + \sum_{n'\neq n} v_{n,n'}\,\psi_{n'} \tag{458}$$

This set is very similar to the set which describes electronic excitations (e.g., the set in Equation 416) except that spacing between k_n^2 and k_{n+1}^2 is relatively small and $v_{n,n}$ do not change appreciably from one channel to another (not as much as in Figure 33). In fact, as already noted in the chapter on semiclassical theory, the number of channels increases in the limit $h \rightarrow 0$.

The choice of potential V(r, ρ) in Equation 451 depends very much on circumstances. If the kinematic mode of energy transfer is dominant then V(r, ρ) will depend essentially on the relative separation of the incoming atom and the nearest atom in the molecule. If the mass of incoming atom is M, the mass of the nearest atom in the diatomic molecule is m_2, and the mass of the other atom is m_1, then in such a case V(r, ρ) will be a function of the form $V\left(\frac{m_1 + m_2 + M}{m_1 + m_2}r - \frac{m_1}{m_1 + m_2}\rho\right)$. Since the matrix elements $v_{n,n}(r)$ are in WKB approximation

$$v_{n,n}(r) \sim \int_{\rho_1}^{\rho_2}\frac{1}{p}V\,d\rho \Big/ \int_{\rho_1}^{\rho_2}\frac{d\rho}{p} \tag{459}$$

where p is the momentum of the oscillator, and since for low vibrational states of the molecule ρ_1 and ρ_2 are close to each other, the elements $v_{n,n}$ have the approximate functional form $v_{n,n}(ar - b\rho_0)$, where ρ_0 is the equilibrium distance of the molecule. For higher n states $v_{n,n}$ may not have such a form, but we notice that in general $v_{n,n}$ do not cross. This is contrary to the case in electronic excitations, where potentials should cross if there is going to be energy exchange between electronic states in semiclassical limit. However, in vibrational energy transfer the separation between channels is relatively small so that in the adiabatic Hamiltonian the coupling elements between channels are large, thus compensating for the fact that channel potentials do not cross.

We can relatively easily make an estimate of the conditions in which long-lived states are formed when kinematic energy transfer is dominant. By definition, these states are formed when the energy transfer from the translational motion into the vibrations of the molecule is so large that the atom after collision stays bound to the molecule. Let us assume that before and during collision the molecule is not vibrating until the moment when the atom hits the nearest atom of the molecule. Just before the impact the relative translational energy of the atom and molecule equals the difference between the initial collision energy E and the lowest value of the potential energy. From this we can calculate the initial relative velocity of the atom and molecule, and if we assume that atoms behave as billiard balls we can obtain velocities of all atoms after collision. We find that the final relative velocity of the atom with respect to the molecule, which is now vibrating, is smaller in magnitude than the initial relative velocity, by a factor

$$Q = \frac{m_2(m_1 + m_2) + M(m_2 - m_1)}{(m_1 + m_2)(m_2 + M)} \qquad (460)$$

The difference in the magnitude of relative velocity is attributed to energy loss, and if this loss is greater than the total energy of collision E, the atom will not be able to escape from the molecule. We find that this happens when

$$1 - Q^2 = \frac{4\,M\,m_1m_2(m_1 + m_2 + M)}{(m_1 + m_2)^2\,(m_2 + M)^2} > \frac{1}{1 + \frac{V_0}{E}} \qquad (461)$$

where V_0 is the absolute value of the lowest point of potential. This condition is approximate and is only used as a guidence if there is a chance of formation of long-lived states. Deviations from this rule may come because we did not take into account vibration of the molecule (even in its ground state) or dynamic effects.

The mass effect on transition probabilities can be explicitly obtained from the dependence of V on r and ρ. We can make the expansion

$$V(ar - b\rho) \sim V(r) + \frac{Mr - m_1\rho}{m_1 + m_2} V'(r) \qquad (462)$$

from which we obtain approximate mass dependence of the coupling terms in Equation 458 and from that the mass dependence of the transition probabilities. In general, this approximation will be valid if m_2 is large compared to m_1 and M. However, very often the potential is not written in the form which explicitly exhibits the distance between the incoming atom and one atom in the molecule. In such a case, the mass effect is not so obvious, but it is nevertheless there. The kinematic effects can be sometimes neglected compared to the dynamic effects. The potential which neglects the kinematic effects is a function of r and ρ, where these two variables are entirely independent of each other. One form of such potential is a product $V(r, \rho) = V_1(r)V_2(\rho)$, which is a generalized form of ion-dipol interaction. $V_2(\rho)$ plays the role of dipol moment under the influence of external charge (hence, we can also write $V(\rho) \sim \alpha + \beta\,\rho$) while $V_1(r)$ is the equivalent of the ion-dipol potential, of course with the appropriate short range repulsive term.

The emphasis on one of these two modes of energy transfer will be also reflected in the coupling matrix of the set of Equations 458. In order to see this let us assume that V(r, ρ) is infinite below a certain value of r and ρ. The matrix elements $v_{n,n'}$ in Equation 458 are essentially given by

$$v_{n,n'} \sim \int_{\rho_1}^{\rho_2} \frac{\cos(\Delta)}{\sqrt{p_n\, p_{n'}}} V(r, \rho)\, d\rho \tag{463}$$

where ρ_1 and ρ_2 are turning points which are chosen in such a way that both p_n and $p_{n'}$ are real. Δ is the difference of two phase integrals for the states n and n′. In general, when n and n′ increase, the turning point ρ_1 acquires smaller value while ρ_2 becomes larger. If V(r, ρ) emphasizes the kinematic effects, then for small n and n′, and some fixed r, the elements $v_{n,n'}(r)$ will be finite, because ar − b ρ_1 in Equation 462 is greater than the radius of the infinite wall. However, when n and n′ are large the elements $v_{n,n'}(r)$ will be infinite, because the integral in Equation 463 includes also a portion of V(r, ρ) which is infinite. Therefore, for a given r, the coupling matrix $v_{n,n'}(r)$ contains both large and small elements. On the other hand, if V(r, ρ) emphasizes the dynamic effects, the integral in Equation 463 is finite for all n and n′, which indicates that there is essentially no large vibrational coupling of states. This only reflects the fact that kinematic energy transfer is much more efficient than the dynamic mode.

Likewise the approach to solving the set of Equations 458 will be different in the two cases. While it is expected that the perturbation theory could be applied when the dynamic mode is dominant, such an approach for the kinematic mode of energy transfer is only possible in relatively few cases. Let us first consider the case of the dynamic mode of energy transfer. Without losing too much on generality, we can assume that the molecule is a harmonic oscillator, in which case the functions φ_n in Equation 452 are

$$\varphi_n(Q) = (2\,\kappa m_{12})^{1/4} (\sqrt{\pi}\, 2^n n!)^{-1/2} e^{-1/2Q^2} H_n(Q) \tag{464}$$

where $H_n(Q)$ are Hermit polynomials and $Q = (2\kappa m_{12})^{1/4}(\rho - \rho_0)$. For the potential V_M in Equation 451 we have assumed the form $V_M = (\rho - \rho_0)^2$. The molecule has reduced mass m_{12}. The matrix elements $v_{n,n'}$ are now scaled as

$$v_{n,n'} = 2\mu(2\kappa m_{12})^{-1/4} \int_{-\infty}^{\infty} dQ\, \varphi_n(Q)\, V[r, \rho_0 + (2\kappa m_{12})^{-1/4}Q]\, \varphi_{n'}(Q') \tag{465}$$

where again we have explicitly taken into account the mass effect, but now in terms of the reduced mass of the molecule. Most of the contribution to the integral in Equation 465 comes from the interval between the two turning points of the kinetic energy of the harmonic oscillator, which are given by $Q_{1,2} = \pm\sqrt{2n + 1}$. Therefore, for those states for which $|Q_{1,2}|\,(2\kappa m_{12})^{-1/4}$ is small, we can write approximately

$$V(r, \rho) \sim V(r, \rho_0) + (2\kappa m_{12})^{-1/4}\, Q\, V_{(1)}(r, \rho_0) + \frac{(2\kappa m_{12})^{-1/2}}{2} \cdot Q^2\, V_{(2)}(r, \rho_0) \tag{466}$$

so that the diagonal elements $v_{n,n}$ are

$$v_{n,n} \sim \left[V(r, \rho_0) + \frac{n + 1/2}{2\sqrt{2\kappa m_{12}}} V_{(2)}(r, \rho_0)\right] 2\mu \tag{467}$$

while the first off-diagonal ones are

$$v_{n,n+1} = v_{n+1,n} \sim -2\mu(2\kappa m_{12})^{-1/4} \sqrt{n/2 + 1/2}\, V_{(1)}(r, \rho_0) \tag{468}$$

In general, it can be shown that the off-diagonal elements have an order of magnitude estimate $v_{n,n+m} \sim (\kappa m_{12})^{-m/4}$, and therefore $v_{n,m}$ can be regarded as expansion in powers of $m_{12}^{-1/4}$. This means that for large m_{12} the diabatic expansion is the most convenient way of describing the dynamic mode of transitions. On the other hand, when m_{12} is small the off-diagonal elements $v_{n,m}$ are large, and one should consider using adiabatic expansion for describing this kind of energy transfer. In this expansion one solves the equation

$$[T_\rho + V_M + V(r, \rho)]\, \varphi_n(\rho;\ r) = E_n\, \varphi_n(\rho;\ r) \tag{469}$$

for each r and expand ψ in Equation 451 in the series

$$\psi = \sum_n \varphi_n(\rho;\ r)\, \omega_n(r) \tag{470}$$

where $\omega_n(r)$ satisfy the set of Equations 339. In such a case the elements of v in Equation 339 are given by

$$v_{m,n} = -v_{n,m} = -2 \int \varphi_m \frac{\partial}{\partial r} \varphi_n\, d\rho \tag{471}$$

We can simplify discussion if it is assumed that $V(r, \rho)$ has the expansion in Equation 466, and only the linear term is retained. The only nonvanishing elements of $v_{m,n}$ are in such a case

$$v_{m,n} = (2\kappa m_{12})^{1/4} \frac{V'_{(1)}}{2\kappa} \left(\sqrt{2n}\, \delta_{m,n-1} - \sqrt{2(n+1)}\, \delta_{m,n+1}\right) \tag{472}$$

which are small when m_{12} is small. Therefore, adiabatic Hamiltonian is a more appropriate description of collision when m_{12} is small or when κ is large, but this is exactly the case when the frequency of the oscillator ω_0 is high ($\omega_0 = [2\kappa/m_{12}])^{1/2}$). Physically, the two limits of m_{12} correspond to two different kinds of energy transfer, which are both very inefficient, and therefore transition probabilities in both cases are small. When m_{12} is large the force which acts between the two atoms in the molecule is not sufficient to pull them apart in a short interval of time during collision, and therefore the molecule is unable to absorb large amounts of energy. When m_{12} is small, the molecule oscillates so rapidly that at any given moment of collision time it is in a stationary state, and therefore after collision it has the same amplitude as before collision. In this case, there is also no energy transfer. The largest energy transfer occurs for some intermediate case which will be when the collision time is about one quarter of the oscillation time of the molecule. It should be emphasized that what we have derived applies when only the dynamic mode of energy transfer is dominant.

We can now analyze the complex angular momentum poles for the case of vibrational energy transfer. If we replace v by $v_0 + \epsilon v'$, where v' contains only the off-diagonal elements of v, then each pole $\lambda_\mu = l_\mu + {}^1/_2$ is a function of ϵ. In particular, for $\epsilon = 0$ of these poles can be found with relative ease, because the set of Equations 458 reduces, in such a case, to a set of independent one channel equations. The set of poles λ_μ is therefore split into subsets of poles, according to the channel in which they belong when $\epsilon = 0$. Therefore, in order to distinguish those various poles, we will give them another label n, which indicates that the pole $\lambda_{\mu,n}(\epsilon)$ is the μth pole of the channel n, i.e., $\lambda_{\mu,n}(0)$ solves the equation

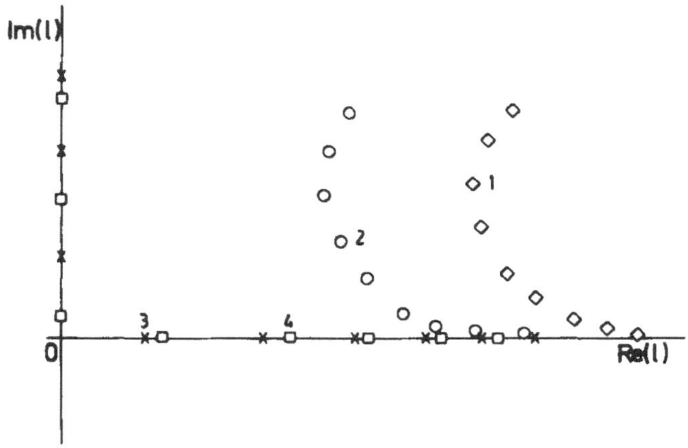

FIGURE 38. Typical distribution of unperturbed poles for vibrational energy transfer. Numbers indicate channels to which particular poles belong. Channels 1 and 2 are open while 3 and 4 are closed.

$$\psi_n'' = \left(v_{n,n} + \frac{\lambda_{\mu,n}^2 - 1/4}{r^2} - k_n^2\right)\psi_n \tag{473}$$

with the boundary condition $\psi_n \sim e^{ik_n r}$. This is an ordinary, single channel problem, which has been discussed *in extenso* in the previous chapter. Thus, for $k_n^2 > 0$ and $v_{n,n}$ typical of atom-atom potential, the poles $\lambda_{\mu,n}(0)$ are complex and, in general, non-ZAM poles, because $v_{n,n}$ usually does not have a barrier. Their typical distribution is shown in Figure 23. For each open channel there will be a separate set of poles $\lambda_{\mu,n}(0)$, and each set will be shifted with respect to the other because k_n^2 differs from channel to channel.

For closed channels, the poles $\lambda_{\mu,n}(0)$ are either real or imaginary because they give value of the centrifugal potential $(\lambda^2 - {}^1/_4)/r^2$ for which a bound state of energy $k^2{}_n$ is possible in potential $v_{n,n}$. Properties of such poles were discussed in Section of Chapter 5. Therefore, we can say that the set $\lambda_\mu(0)$ contains complex, real, and imaginary poles, depending on the channel energy of a particular subset. A typical distribution of poles $\lambda_\mu(0)$ for a four-channel case is shown in Figure 38, where we indicate the subset to which a particular pole belongs.

We notice that closed channel poles are predominantly of ZAM type, while open channel poles are of the type shown in Figure 23. Of course, this is not always the case, but we show a typical circumstance. It is obvious that in a many-channel case, the complex λ-plane will be infested with different types of poles, but most of them will be on the real λ-axis. Distribution of poles could be made more transparent by assuming that $v_{n,n}$ do not change much from channel to channel. In such a case Equation 473 gives poles for potential $v_{11} = v_{22} = \ldots = v_{n,n}$ at different collision energies. If it is assumed that for λ_μ^2 this dependence is linear, then the equation $\lambda_\mu^2 = C_\mu (k^2 - k_0^2) + \lambda_{\mu 0}^2$ gives the poles $\lambda_{\mu,n}(0)$, when k is replaced by k_n^2. The index μ refers to the μth bound state of the potential $v_{1,1}$ so that λ_0^μ can be obtained from Equation 352, when $k^2 = k_0^2$. In general, these lines are not parallel, since C_μ is different for different bound states. From the relationship in Equation 355, we find that C_μ increases for increasing μ, and therefore the lines λ_μ^2 are steeper for higher bound states than for the lower ones. This would imply that all these lines cross at certain points, which is not the case since these trajectories are not exactly straight lines. A typical set of such trajectories is shown in Figure 39.

From Figure 39 we can now easily find the unperturbed poles $\lambda_{\mu,n}(0)$. For channel energy k_n^2 they are given at the crossing of the lines λ_μ^2 with the line $k^2 = k_n^2$ (negative λ_μ^2 means

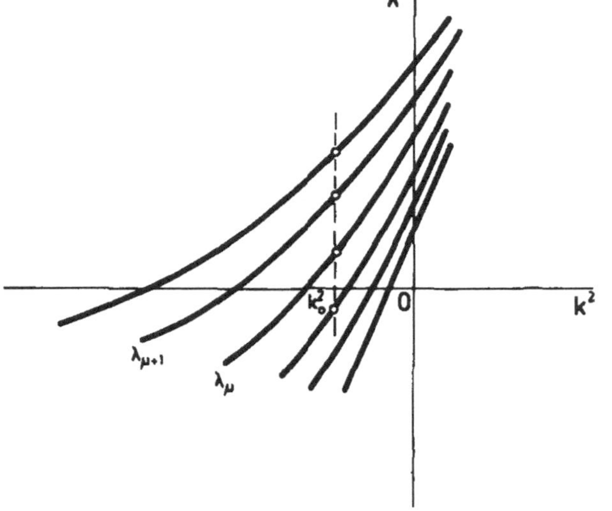

FIGURE 39. Functional form of λ^2 vs. k^2 for bound state poles in vibrational energy transfer.

that the appropriate pole λ_μ is imaginary). In this way we can systematically account for all poles of a given scattering problem.

When perturbation between channels is now included, all poles become complex (except those which represent bound states embedded in continuum) i.e., real poles acquire a imaginary part and imaginary poles acquire a real part, while complex poles move to another place in complex λ-plane. When ϵ is small, movement of poles can be followed by using perturbation theory, from which it follows that $\lambda_{\mu,n}(0) \sim \lambda_{\mu,n}(0) + \frac{\epsilon^2}{2} \lambda^{(2)}_{\mu,n}$, except when two or more poles are nearly degenerate for $\epsilon = 0$. Near degeneracy is accidental when $\epsilon \sim 0$, meaning that although there are many poles they are still sufficiently separated so that ordinary perturbation theory can be applied. The perturbation coefficient $\lambda^{(2)}_{\mu,n}$ is given by Equation 329, which is a sum over integrals between the state n and the other interacting states. If we consider only poles $\lambda_{\mu,n}(0)$ from the closed channels, and if we assume that $v_{n,n}$ are nearly equal, then from the WKB approximation and Equation 436 we obtain an estimate

$$\lambda^{(2)}_{\mu,n}(S) \sim - \frac{1}{4\lambda_{\mu,n}(0)\, A^2_\mu} \int_{r_1}^{r_2} \frac{v^2_{n,s}\, dr}{p_s p_n (p_s - p_n)} \tag{474}$$

so that $\lambda^{(2)}_{\mu,n} = \Sigma\, \lambda^{(2)}_{\mu,n}(s)$. r_1 and r_2 are turning points of p^2_n. We notice that $\lambda^{(2)}_{\mu,n}(s)$ is real for all s; therefore, the largest contribution of the integrals in Equation 329 produces only the shift of poles. The imaginary part can be obtained in a way similar to Equation 423, hence

$$\mathrm{Im}[\lambda^{(2)}_{\mu,n}(s)] = \frac{1}{\lambda_{\mu,n}(0)\, A^2_\mu} \left(\int_{r_1}^{r_2} \varphi_n\, V_{n,s}\, \varphi_s\, dr \right)^2 \tag{475}$$

where now index s refers only to open channels.

In general, $\mathrm{Im}[\lambda^{(2)}_{\mu,n}]$ is smaller than the partial shift in Equation 474, but the overall shift $\mathrm{Re}[\lambda^{(2)}_{\mu,n}]$ may not be so large because when $\lambda^{(2)}_{\mu,n}(s)$ are summed over all s then some terms will cancel, e.g., the contributions $\lambda^{(2)}_{\mu,n}$ (n + 1) and $\lambda^{(2)}_{\mu,n}$ (n − 1) have opposite signs.

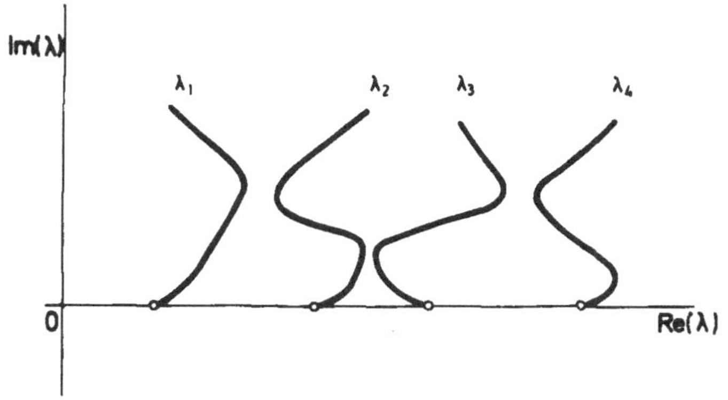

FIGURE 40. Dependence of poles on perturbation. For small perturbation their path is nearly parabolic, while for larger perturbation they behave as degenerate poles.

When ϵ is further increased the function $\lambda_{\mu,n}(\epsilon)$ will start to deviate from quadratic function, and at some point the coupling between channels will become so strong that neighboring poles may approach each other, with a tendency that their paths cross. However, at this point the nondegenerate perturbation theory breaks down, and one should apply nearly degenerate perturbation theory. It follows from this theory that the path of poles which approach each other will not cross, and instead an exchange effect occurs, very similar to an avoided crossing effect of potential curves in atom-atom systems. In Figure 40 we show a possible ϵ-dependence of a few poles.

We notice that at the point when poles start interacting with each other, the analysis which is based on perturbation arguments breaks down. Mixing of poles will be so strong, when there are many of them, that we can talk about randomization of states among those which are accessible by energy transfer. Further increase of ϵ leads to even more mixing, although this may not always be true. For example, in purely dynamic energy transfer we can take for the coupling constant in Equation 468 the value $m_{1/2}^{-1/4}$. As the mass m_{12} is decreased there is more energy transfer into the molecule; however, beyond certain m_{12} the energy transfer becomes less efficient, and in fact it goes to zero for $m_{12} \rightarrow 0$, which is obtained from Equation 472. Therefore, not always stronger coupling means more mixing of states. In this particular example, mixing of states, if it occurs, happens for some intermediate values of m_{12}. On the other hand, in purely kinematic energy transfer, stronger coupling usually leads to more energy transfer, and by stronger coupling, in this case we mean, for example, large value of the ratio $M/(m_1 + m_2)$. Formally, the difference between these two kinds of energy transfer can be seen, as explained earlier, in the structure of $v_{m,n}$; for dynamic energy transfer the magnitude of $v_{m,n}$ is progressively smaller for a large difference $m - n$, while for kinematic energy transfer, these elements are increasing. Therefore, it is expected that mixing of the poles in the kinematic energy transfer will be strong even for relatively small coupling, while in the dynamic energy transfer this mixing will only involve poles from channels which are close to the open channels.

However, once mixing of poles starts their energy dependence becomes erratic, and no longer can we expect smooth energy behavior of poles. We have already seen in the two channel case, discussed in the previous chapter, that energy dependence of poles can be quite complicated (see Figure 36), and when there are many more channels, this dependence can be very complex. One of the consequences of such energy behavior of poles, apart from the fact that their energy extrapolation becomes very difficult, is that the structure of time delay gets very complicated. This is because time delay in collisions is approximately proportional to the energy derivative of the real part of λ_μ. However, there is another important consequence on time delay when there is mixing of poles. The energy derivative of poles is given by Equation 298, and the integral in the denominator is given by generalization of Equation D15 to the multichannel case. We have

$$i\, J^+ K \frac{\partial}{\partial \lambda} J^- = -\lambda \int_{r_0}^{\infty} \tilde{\psi}\, \psi\, dr/r^2 \tag{476}$$

and if we assume that two or more poles are degenerate, then $\partial/\partial\lambda\ J^- = 0$, which implies that the corresponding time delay is infinite. Since poles do not become degenerate, the derivative $\partial/\partial\lambda\ J^-$ is not zero, but is very small, meaning that the time delay is large. Therefore, mixing of poles also initiates formation of very long-lived states, in fact much longer lived than the states which we discussed in elastic collisions and which correspond to ZAM poles. It would be tempting to assume that due to mixing, behavior of poles is random. However, we cannot give any apparent reason that this happens since not many of the properties of these poles are known.

In the presence of so many poles, the scattering amplitude becomes more difficult to analyze, and especially when mixing of poles sets in. In the weak coupling case we can calculate residues from perturbation theory, given by Equation 284. When the unperturbed pole belongs to closed channel n, then for any open channel s

$$(\beta_\mu)_{s,s} = -\frac{e^{i\pi\lambda_{\mu,n}}}{4\, \lambda_{\mu,n}(0)\, A_\mu^2 (j_s^-)^2} \left(\int_{r_0}^{\infty} \varphi_s v_{s,n}\, \varphi_{n,\mu} dr \right)^2 \tag{477}$$

If this expression is compared with Equation 475, we notice that elastic residue is proportional to the partial imaginary part of the pole in the open channel s. There is an interesting consequence of this fact. The partial width of a pole $\lambda_{\mu,n}$ can be zero for a particular open channel s. In such a case resonances, which are described by this pole, will not be observed in all inelastic transitions involving open channel s. In fact, some resonances may not be observed at all, which is the case with bound states embedded in continuum. Properties of these states were discussed in the previous chapter.

Scattering amplitude in backward space is given by Equation 273, and a typical element from the sum over the poles is

$$f_{s,t}^{(\mu,n)} = -\pi\, i\, \frac{\lambda_{\mu,n} (\beta_\mu)_{s,s}^{1/2} (\beta_\mu)_{t,t}^{1/2}}{K_t \cos(\pi\, \lambda_{\mu,n})} P_{\lambda_\mu - 1/2}(-\cos\theta) \tag{478}$$

The number of these terms can be very large, and therefore it would appear that any analysis of this part of the scattering amplitude is almost impossible. However, if we write this amplitude keeping only terms which oscillate when μ and n change, then

$$f_{s,t} \sim \sum_{\mu,n} \frac{e^{i\pi\lambda_{\mu,n}}}{\cos(\pi\, \lambda_{\mu,n})} P_{\lambda_{\mu,n} - 1/2}(-\cos\theta)\, \langle s | v_{s,n} | n, \mu \rangle \langle t | v_{t,n} | n, \mu \rangle \tag{479}$$

where we have also retained the matrix elements of v, which may be also oscillatory. Contribution of a single pole, even when it gives rise to a resonance, is without meaning because it is screened by contributions of other poles. However, a collective effect of many poles can give rise to a more general feature of scattering amplitude. Such an effect is observed if for a group of poles the phase of the terms in the sum does not change, e.g., for a cluster of poles which have nearly the same real and imaginary parts. In general, such a condition will be strongly dependent on the interaction between atom and molecule, because for a uniform distribution of poles and if oscillations of matrix elements of $v_{s,n}$ are neglected, the amplitude $f_{s,t}$ will be on average very small.

The background integral in the scattering amplitude of Equation 273 is

$$f_{s,t}^{(B)} = \frac{1}{K_t} \int_0^{\infty} d\lambda\lambda \; S_{s,t}^{(\lambda - 1/2)} \, e^{-i\pi\lambda} \, P_{\lambda - 1/2}(-\cos\theta) \tag{480}$$

where for simplicity we have retained only the integration path along real λ. In order to evaluate this integral we will use a theorem which says that an analytic function can be represented in terms of its poles, plus a remainder which is nowhere singular. The exact way this parametrization is done is not uniquely prescribed, because we know very often additional information about the function which can be directly incorporated in the parametrization, e.g., some symmetry properties. For the S-matrix we choose the following representation

$$S_{s,t}^{(\lambda - 1/2)} = \sum_{\mu,n} \frac{2\lambda_{\mu,n}}{\lambda_{\mu,n}^2 - \lambda_{\mu,n}^{*2}} \frac{\lambda^2 - \lambda_{\mu,n}^{*2}}{\lambda^2 - \lambda_{\mu,n}^2} |e^{-i\pi\lambda_{\mu,n}}(\beta_\mu)_{s,t}| \, e^{i\pi\lambda}\gamma_{s,t}(\lambda) \tag{481}$$

which ensures the symmetry of Equation 246 if $\gamma_{s,t}(-\lambda) = \gamma_{s,t}(\lambda)$. The coefficients are chosen in such a way that $\gamma_{s,t}(\lambda)$ is in the form $e^{i\delta_{s,t}}$ with ambition that $\delta_{s,t}$ is not a rapidly varying function of λ. In deciding about the coefficients we used as guidance Equation 477, and therefore for strong coupling $\gamma_{s,t}$ may be a nonuniform function of λ; however, it is hoped not too much.

$\gamma_{s,t}(\lambda)$ may be obtained from the following arguments. If $\lambda = \lambda_{\mu,n}$ the function $\gamma_{s,t}(\lambda)$ should reproduce corresponding residues, which gives

$$\gamma_{s,t}(\lambda_{\mu,n}) = e^{i\arg(\beta_\mu) - i\pi\mathrm{Re}(\lambda_{\mu,n})} \tag{482}$$

Since by assumption there are many poles in the vicinity of the real λ-axis and $\gamma_{s,t}(\lambda)$ is not singular, all $\gamma_{s,t}(\lambda_{\mu,n})$ constitute a set of points from which we can obtain $\delta_{s,t}(\lambda)$ by interpolation with polynomial. With such a function $\gamma_{s,t}(\lambda)$ we can evaluate the integral

$$I_{\mu,n} = \int_0^{\infty} d\lambda \frac{\lambda}{\lambda^2 - \lambda_{\mu,n}^2} e^{i\delta_{s,t}(\lambda)} P_{\lambda - 1/2}(-\cos\theta) \tag{483}$$

which is entirely analogous to the integral which appeared in the study of ZAM resonances in elastic collisions. The difference is that $\delta_{s,t}(\lambda)$ may not have simple λ behavior as in the case of elastic collisions. The integral in Equation 483 is part of the scattering amplitude in Equation 273, and this part was discussed for the case of single channel collisions and single pole contributions. Therefore, in the case when many poles are present in multichannel collisions, the background scattering amplitude is

$$(f_B)_{s,t} = \frac{1}{K_t} \sum_{\mu,n} \frac{\lambda_{\mu,n}|e^{-i\pi\lambda_{\mu,n}}(\beta_\mu)_{s,t}|}{2\,\mathrm{Re}(\lambda_{\mu,n})\,\mathrm{Im}(\lambda_{\mu,n})} \int_0^{\infty} d\lambda\,\lambda\, e^{i\delta_{s,t}(\lambda)} P_{\lambda - 1/2}(-\cos\theta) + \frac{1}{K_t} \sum_{\mu,n} 2\lambda_{\mu,n}\,|e^{-i\pi\lambda_{\mu,n}}(\beta_\mu)_{s,t}|\, I_{\mu,n} \tag{484}$$

The first term is analogous to the direct reflection term in Equation 401, and its module is essentially determined by the sum of ratios $\{Im[\lambda_{\mu,m}(s)] \cdot Im[\lambda_{\mu,n}(t)]\}^{1/2}/Im(\lambda_{\mu,n})$, where the partial imaginary parts were defined in Equation 475. The exception is when the pole belongs to either channel s or t, i.e., when $\lambda = \lambda_{\mu,s}$ or $\lambda = \lambda_{\mu,t}$. In such a case this rule may not apply when these poles have a large imaginary part. It should be recalled that residues for such poles have nonzero unperturbed value. On the other hand, the magnitude of the second term is mainly determined by the sum of the partial widths of the poles, and in the weak coupling case this term can be neglected compared to the direct reflection term. In a strong coupling case, the situation can easily be the other way around.

We did not discuss the total cross-section, because from the optical theorem we can only get the total cross-section in all open channels, which is not often of much interest. In the next section we will derive a more general form of optical theorem which also includes total cross-section for a particular transition.

B. Inelastic Total Cross-Sections

The disadvantage of the optical theorem for inelastic collisions is that it only gives total cross-section in all open channels; very often it is of interest to have total cross-section for a particular transition. A straightforward way of obtaining such a cross-section would be either from partial wave decomposition or from one of the forms in Equation 264 or 273 of the scattering amplitude, by integrating differential cross-section over the solid angle. We will give here an alternative, which is based on the idea of complex angular momentum.

The inelastic total cross-section is given by

$$\sigma_{n,m} = \frac{\pi}{2k_m^2} \sum_{l=0}^{\infty} (2l + 1)\, S_{n,m}^{(l)}[S^{(l)}]_{n,m}^{-1} \tag{485}$$

which was obtained from Equation 243. We have used in Equation 485 the unitary property of the S-matrix, which is for a general case when l is complex,

$$[S^{(l)}]^{-1} = S^{(l^*)*} \tag{486}$$

If we now apply the transformation in Equation 244 on Equation 485, the inelastic cross-section is

$$\sigma_{n,m} = \frac{\pi}{K_m^2} \sum_{M=-\infty}^{\infty} e^{-iM\pi} \int_0^{\infty} d\lambda\, \lambda\, S_{n,m}^{(\lambda-1/2)}[S^{(\lambda-1/2)}]_{n,m}^{-1}\, e^{2\pi i M\lambda} \tag{487}$$

The function

$$Q_{n,m}^{(\lambda)} = S_{n,m}^{(\lambda-1/2)}[S^{(\lambda-1/2)}]_{n,m}^{-1} \tag{488}$$

is symmetric with respect to the reflection $\lambda \to -\lambda$, and also it is zero on the semicircle $|\lambda| \to \infty$; $|\arg(\lambda)| \leq \pi/2$. Furthermore, it has poles at $\lambda = \lambda_\mu$, but also at $\lambda = \lambda_\mu^*$, which follows from Equation 486. Therefore, we can transform 487 into

$$\sigma_{n,m} = \frac{\pi}{K_m^2} \int_0^{\infty} d\lambda\, \lambda\, Q_{n,m}^{(\lambda)} - \frac{\pi}{2\,K_m^2} \int_{0^+}^{\infty} d\lambda\, \lambda\, Q_{n,m}^{(\lambda)} \frac{e^{+i\pi\lambda}}{\cos(\pi\lambda)}$$
$$- \frac{\pi}{2\,K_m^2} \int_{0^-}^{\infty} d\lambda\, \lambda\, Q_{n,m}^{(\lambda)} \frac{e^{-i\pi\lambda}}{\cos(\pi\lambda)} \tag{489}$$

where the meaning of 0^+, and likewise that of 0^-, was explained in Equation 248. If we designate by $q_{n,m}^{\lambda_\mu}$ the residue of $Q_{n,m}^{\lambda}$ at $\lambda = \lambda_\mu$, then the above expression transforms into

$$\sigma_{n,m} = \frac{\pi}{K_m^2}\int_0^\infty d\lambda\ \lambda\ Q_{n,m}^{(\lambda)} - \frac{\pi^2 i}{K_m^2}\sum_\mu \left[\lambda_\mu\ q_{n,m}^{(\lambda_\mu)} \frac{e^{i\pi\lambda_\mu}}{\cos(\pi\lambda_\mu)} - \text{c.c.}\right] \tag{490}$$

where we have neglected the integral along the imaginary axis. The letters c.c. indicate the complex conjugate of the first term in the bracket. The residues $q_{n,m}^{\lambda_\mu}$ are obtained from Equations 488, 486, and 481, and they are given by

$$q_{n,m}^{(\lambda_\mu)} = (\beta_\mu)_{n,m}\sum_{\nu\neq\mu}\frac{2\ \lambda_\nu^*}{\lambda_\nu^* - \lambda_\nu^2}\frac{\lambda_\mu^2 - \lambda_\nu^2}{\lambda_\mu^2 - \lambda_\nu^{*2}}\left|e^{-i\pi\lambda_\nu}(\beta_\nu)_{n,m}\right| e^{-i\pi\lambda_\nu}\ \gamma_{n,m}^*(\lambda_\nu^*) \tag{491}$$

We have now sufficient data to calculate the inelastic cross-sections. The sum over the residues is quite straightforward once $q_{n,m}^{\lambda_\mu}$ are known, although the sum may contain many terms. The integral, on the other hand, could be transformed into the integral along the imaginary axis, but in this way we would not achieve much since the integrand is not a rapidly decaying function on this axis. Therefore, it should be evaluated as it is, most probably numerically. However, we can give its order of magnitude estimate. From Equation 481 we obtain the most dominant term of the S-matrix, which is

$$S_{n,m}^{(\lambda - 1/2)} \sim \sum_\mu \frac{2\ \lambda_\mu}{\lambda_\mu^2 - \lambda_\mu^{*2}}\left|e^{-i\pi\lambda_\mu}(\beta_\mu)_{n,m}\right| e^{i\pi\lambda}\ \gamma_{n,m}(\lambda) \tag{492}$$

and therefore the integral in Equation 490 is approximately

$$\int_0^\infty d\lambda\ \lambda\ Q_{n,m}^{(\lambda)} \sim \left|\left|\sum_\mu \frac{2\lambda_\mu}{\lambda_\mu^2 - \lambda_\mu^{*2}}\right| e^{-i\pi\lambda_\mu}(\beta_\mu)_{n,m}\right|^2 \int_0^\infty d\ \lambda\lambda\ |\gamma_{n,m}(\lambda)|^2 \tag{493}$$

C. Classical Long-Lived States

We come now to a point when we should ask a question which is in the basis of all our understanding of long-lived states in collisions. However, before doing this we must describe the circumstances. All collision processes which we described, except those where electronic states are involved, can be treated by either the quantum or classical theory of collisions. The change is simple and involves only going from Schrödinger's to Newton's equations of motion. If we describe a particular system by the classical theory, we may or we may not observe long-lived states, in the sense which was described in Chapter 1. There may be various reasons why long-lived states are not formed, e.g., the mass of the incoming atom is so small that energy transfer into the molecule is negligible. On the other hand, in quantum description of collisions we always obtain a multitude of poles, from which we can derive the conclusion that long-lived states are always formed. Quantum long-lived states (that is how we will call those states which we interpret as being long lived on the basis of Schrödinger's equation) are formed for no matter how small a coupling between channels (e.g., for very small mass of incoming atom), which is definitely not the case with classical long-lived states. Very often, the very presence of resonances is attributed, wrongly, to formation of such states. In the extreme case when coupling is very weak, resonances (which originate in closed channels) will have very small width, meaning that very long-lived states are formed, which is contrary to the classical point of view.

The question which we want to ask is how do we know that quantum long-lived states (and here we do not mean resonances) indicate that classical long-lived states also are formed.

In other words, what happens to the poles which is an indication that classical long-lived states also are formed? It is obvious that in order to answer this question we must study the behavior of poles as a function of coupling between channels. Before answering this question we should understand what is the meaning of complex angular momentum poles in the classical theory of collisions. These poles are principally connected with the boundary condition of the wave function; for such a pole the regular wave function has only the e^{ikr} component for large r. However, the wave function can be reconstructed from classical trajectories (see the chapter on semiclassical theory), and this link can be used to analyze poles in classical theory. From such analysis we will be able to make distinction between the classical and quantum long-lived states. If angular momentum is complex in the classical theory of collisions, then the trajectory is also complex, which means that we must know the potential for complex coordinates. We will assume that we know this, and furthermore it will be assumed that the potential is analytic in a sufficiently large domain around the real trajectory. The time variable is kept real, and for this case we will call the phase space of all complex trajectories the classical phase space. On the other hand, we will call the phase space which is accessible if time is complex the nonclassical phase space (e.g., tunneling effect was explained in Chapter 1 when time was allowed to be complex).

Radial momentum squared is given by

$$p^2 = k^2 - \frac{l^2}{r^2} - V \tag{494}$$

where p is the operator in quantum theory and p = mv in classical theory. If time t is renormalized by $t \rightarrow t/m$, then the classical equation of motion is

$$\frac{dv}{dt} = \frac{l^2}{r^3} - \frac{1}{2}\frac{dV}{dr} \tag{495}$$

We will assume that the potential is of the form shown in Figure 16 in which case Equation 494 may have either one or three real turning points for real l (see Figure 20). Let us assume for the moment that Equation 494 has three real turning points A, B, and C, shown in Figure 20b. When l is made complex, with small and positive imaginary parts, then A, B, and C become complex, and their position is shown in Figure 41a. A typical classical trajectory is also shown in Figure 41a, by a solid line.

As we notice, classical trajectory for complex l does not come near the point A, and what is more important, it does not go around it. For this trajectory we can say that the space around A is a classically inaccessible region. We can now calculate the phase and the normalization factor of the scattering wave function, which corresponds to the trajectory in Figure 41a. The rules were given in Chapter 3, and if they are formally applied to our case, we obtain that the phase of the outgoing component of the wave function is a line integral of the form

$$\delta \sim \int_L v\, dr \tag{496}$$

where L designates the classical trajectory. Since the potential is analytic in the domain around Re(r), the integral in Equation 496 can be extended to any other path which does not encircle A (e.g., the broken line path shown in Figure 41a). The path cannot encircle A because v in Equation 496 has square root branch point there. Therefore, the regular wave function will have the asymptotic form.

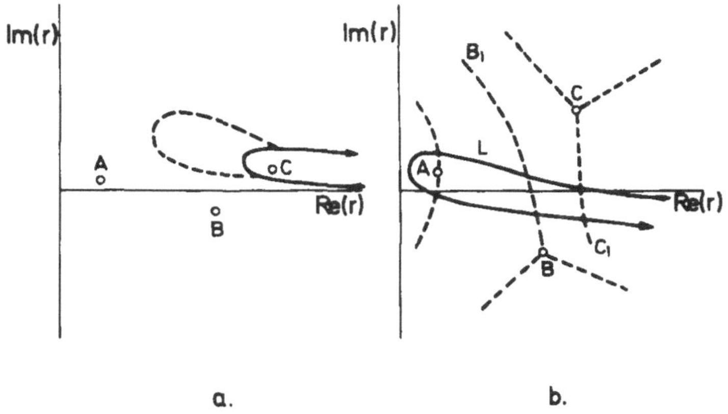

FIGURE 41. (a) Three complex turning points for the potential in Figure 20b when angular momentum is complex Complex classical trajectory is shown by a solid line. (b) Complex turning points for the potential in Figure 20a when l is complex. Broken lines show Stoke's lines, and L is classical trajectory

$$\psi \sim e^{-ikr} + e^{i\delta} \qquad (497)$$

which can only have the outgoing component if $e^{i\delta}$ is infinite. However, we see from Equation 496 that this is not possible, hence we conclude that when v^2 has three real turning points for real angular momentum, there are no complex angular momentum poles with small imaginary part. This is contrary to what was found earlier in Equation 353, where it was shown that just for this case there are poles with small imaginary part. The only conclusion is that such poles are not classical. Indeed, if time is allowed to be complex, trajectory can reach A and go around it. In fact, this was demonstrated in Chapter 1 when tunneling resonances were discussed.

Our analysis was too crude to be of any use for a more general case. The main objection in derivation of Equation 497 is that we have assumed that $e^{i\delta}$ is the dominant solution along the entire trajectory L and that on the same path L the solution $e^{-i\delta}$ can be neglected. This is not true because there are regions of the complex r plane when, for example, the phase integral in Equation 496 acquires only positive imaginary part, in which case $e^{i\delta}$ will decrease and $e^{-i\delta}$ increase. Therefore, the component $e^{-i\delta}$, which was until then negligible (subdominant), suddenly becomes more important than $e^{i\delta}$, which becomes small. The line which separates such regions is also called the Stoke's line. As it turns out, each turning point of v^2 has its own Stoke's lines, altogether three for each turning point if the turning point is first order zero of v^2. The rules which tell what happens to dominant and subdominant components of the wave function when trajectory crosses Stoke's lines were analyzed in great detail.[64,65] When they are applied to the case shown in Figure 41a, we indeed obtain the solution in Equation 497.

For the turning point A the Stoke's lines are defined as

$$\mathrm{Re}\left(\int_A^r v\, dr'\right) = 0 \qquad (498)$$

which are shown in Figure 41b by broken lines. Distribution of turning points A, B, and C differs from that in Figure 41a because now there is a classical trajectory which encircles A (shown by solid line L). If l is real, this case would correspond to the single turning point problem in Figure 20a, where B and C are complex. We notice now that the phase integral crosses several Stoke's lines, and therefore single solution of the form $e^{i\delta}$ will not be accurate

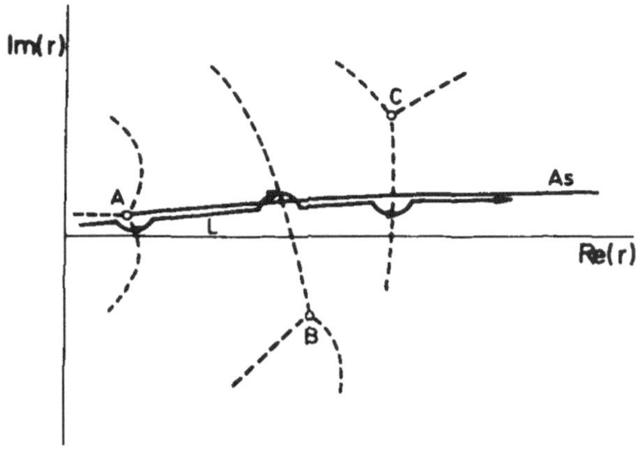

FIGURE 42. The integration path (L) for the WKB rules for the problem in Figure 41b. The anti-Stoke's line is designated by A_s.

along the entire path L. A more accurate solution is obtained if the path L is changed to L_1 (which is allowed because the phase integral is independent of the trajectory, provided that the new path does not encircle an additional turning point), when on crossing the Stoke's line C, the solution $e^{i\delta}$ acquires another component of the form $\exp(-i \int_c^r v dr')$. The rules which describe how to construct the wave function along a trajectory are also known as the WKB rules.[64] However, the WKB rules are usually formulated without reference to trajectories, but rather to some integration paths. In our case we make the restriction that the path coincides with the classical trajectory or any other equivalent path. Thus, for example, the WKB solution, which corresponds to the trajectory in Figure 41b is obtained by using the WKB connection rules along the path in Figure 42. The solid line as in Figure 42 designates the anti-Stoke's line, which is defined as the line along which the imaginary part of the phase integral is zero. Along the path prior to point A the WKB solution is proportional to $\exp(\int_A^r |v| \, dr')$, which is a subdominant component in this region. This component connects to

$$\psi \sim e^{i\int_A^r v dr'} + i\left(e^{i\int_A^c v dr} + e^{i\int_c^A v dr}\right) e^{i\int_r^c v dr'} \tag{499}$$

in the region far away to the right. We have neglected the term $\exp(i \int_B^A v \, dr)$. The wave function will have only the component e^{ikr} if

$$e^{2i\int_A^c v dr} = -1 \tag{500}$$

or

$$\int_A^c v \, dr = \pi(n + 1/2) \tag{501}$$

which is generalization of Equation 352 to the case of complex turning points. However, the difference between Equations 501 and 352 is that the former is an equation for "classical" poles while the latter is an equation for "quantum" poles. This distinction is based on our previous definition of quantum and classical long-lived states.

From the previous analysis we can derive behavior of classical poles as a function of collision energy or a perturbation parameter, such as the well depth of the potential. Let us look at a particular quantum pole, which is approximately given by Equation 353. We have seen that this pole will not appear as a classical pole, and therefore we can say that at this collision energy there is no classical long-lived state which corresponds to this pole. The quantum pole in Equation 353 moves as the collision energy is increased, and at a certain point the classical pole will appear. This happens when two real turning points B and C coalesce and become complex. In such a case we can say that the classical long-lived state appears, which corresponds to this pole. Interpretation of this long-lived state was given in Chapter 5, Section II.B. Energy dependence of the quantum pole was shown in Figure 21, which we can now interpret in terms of classical poles. Below energy $k^2 = k_{th}^2$ the pole is nonclassical, but for $k^2 > k^2_{th}$ the pole is classical. In other words, there will be no classical pole for $k^2 < k_{th}^2$, but for $k^2 > k_{th}^2$ it will appear, and its position will more or less coincide with that on Figure 21, Likewise, we can interpret poles in Figure 23. The poles l_μ, $l_{\mu+1}$, . . . are all classical while $l_{\mu-1}$, $l_{\mu-2}$, . . . are all nonclassical. From this we also conclude that all effects which are described by former poles are classical, while the effects described by the latter are nonclassical.

Similar arguments can be applied to inelastic collisions. We will consider atom-molecule collision, when only vibrations of the molecule are taken into account. If the molecule is treated as a harmonic oscillator, then the Hamiltonian of the entire system is

$$H = \omega n + \frac{1}{2M} P^2 + V_l(\rho_0 + \sqrt{n} \cos q, R) \tag{502}$$

where we use canonical variables n and q, n the energy of the oscillator and q its phase. The index l of the potential means that V_l also includes the centrifugal term.

The classical equations of motion are now

$$\mathring{R} = \frac{1}{M} P; \quad \mathring{P} = -\frac{\partial V_l}{\partial R}$$

$$\mathring{q} = \omega + \frac{\cos q}{2\sqrt{n}} \frac{\partial V_l}{\partial \rho}; \quad \mathring{n} = \sqrt{n} \sin q \frac{\partial V_l}{\partial \rho} \tag{503}$$

For simplicity it will be assumed that l is real and that only kinematic energy transfer is considered. In such a case V_l is a function of $aR - b\rho$, except the centrifugal term. The coefficients a and b are related to masses of atoms, and when the mass of the incoming atom is small the coefficient b is also small. In such a case $\partial V_l/\partial \rho$ is small, and the last two equations of Equation 503 are

$$\mathring{q} \sim \omega; \quad \mathring{n} \sim 0 \tag{504}$$

which indicates that there is very little energy transfer into the molecule.

The turning points of P^2 can be obtained from Equation 502, but now they are functions of n and q. We distinguish three cases; when there is only one real turning point of P^2, when there are two, and when there are three. Initially, when the atom is away from the molecule, P^2 will have one real turning point, and the other two are complex. As the atom approaches the molecule, n and q will change, and so will the position of the turning points. When n acquires a large value, two complex turning points may become real, in which case the atom will move in an effective potential similar to that shown in Figure 20b. If the atom is between A and B at the moment of large n change, then it will be trapped in this well until another sufficiently large lowering of n, when B and C become complex and the atom can escape beyond the reach of the potential. For even larger changes of n, the turning point

C can go to infinity, in which case the atom will be trapped between A and B, as in a bound state. Again, until there is a sufficiently large lowering of n, the atom will stay there without being able to escape from the potential.

That was a qualitative description of formation of long-lived states in such collisions, which will help in understanding the difference between the quantum and classical such states. As we mentioned in Equation 504, for very weak coupling there is very little change in n, and therefore the turning points B and C will only slightly change their positions. In such a case there is no chance for formation of classical long-lived states; however, quantum long-lived states can be formed. Let us assume that trajectory reached the turning point A, and that at this moment $q = j\pi; j = 0, \pm 1, \pm 2, \ldots$ From now on the time increments are along the positive imaginary axis. If we make transformation $dt \rightarrow idt$, $P \rightarrow iP$, and $q \rightarrow j\pi + iq$, then the set of Equations 503 becomes

$$\mathring{R} = -\frac{1}{M}P; \quad \mathring{P} = -\frac{\partial V_1}{\partial R}$$

$$\mathring{q} = \omega + (-1)^j \frac{\text{ch}\, q}{2\sqrt{n}} \frac{\partial V_1}{\partial \rho}; \quad \mathring{n} = -(-1)^j \sqrt{n}\, \text{sh}\, q \frac{\partial V_1}{\partial \rho} \tag{505}$$

which has real solution for R and n. When j is even, and for our particular choice of energy transfer, n decreases and, therefore, the turning points B and C will not become real. On the other hand, when j is odd, n increases at least for a period of time, after the change to imaginary time increase. In such a case it is possible that B and C become real, and when this happens one changes to real time increase. The system will be then left in a bound state. However, the change to real time increase is only possible when $q = 0$ and $P = 0$. In order to match this condition we have two parameters at our disposal; q and l.

In the exercise we have shown that no classical long-lived states are formed in low energy transfer collisions, but by a suitable choice of complex time increment there is a possibility of formation of nonclassical long-lived states. Therefore, the poles which correspond to these states we will call nonclassical. When coupling ϵ is now increased, e.g., the mass of the incoming atom is increased or any other parameter which increases energy transfer, the set of Equations 503 may have a solution for which the points B and C become real, in which case we will observe formation of classical long-lived states. In principle, we will be able now to calculate appropriate classical poles. However, for this particular ϵ there may be other long-lived states which are not classical. This can be checked by showing that the set of Equations 505 has a solution which is a long-lived state. Therefore, for a given ϵ there are classical and nonclassical poles, and in general, when ϵ is increased the number of nonclassical poles decreases. We can now deduce qualitative behavior of poles as a function of coupling. For very weak coupling, all poles are quantum, and in principle, their behavior can be described by the perturbation theory. For this coupling there will be no classical poles. When ϵ is increased we will notice that some of the classical poles will appear for $\epsilon = \epsilon_0$. The functional dependence of classical poles on ϵ around $\epsilon = \epsilon_0$ is not known, but it is likely that ϵ_0 is a branch point of the square root type. In Figure 43 we show by a solid line a typical functional dependence of a classical pole on ϵ. Functional dependence of a quantum pole on ϵ is shown by a broken line.

We notice that the quantum pole has two different kinds of behavior, and the turning point is $\epsilon = \epsilon_0$. In the interval $0 < \epsilon < \epsilon_0$ the pole is a smoothly behaving function of ϵ, which has expansion in the power series of ϵ, and therefore it can be obtained from perturbation theory. On the other hand, for $\epsilon > \epsilon_0$ the quantum pole has properties like that of the classical pole, which can be, in principle, reproduced from classical equations of motion. In the transition region around $\epsilon = \epsilon_0$, the quantum pole is very sensitive to variations in

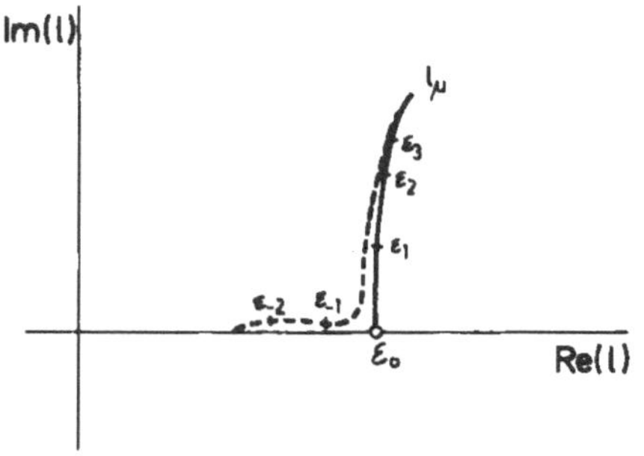

FIGURE 43. Qualitative perturbation dependence of classical (solid line) and quantum (broken line) Regge poles. ϵ_n designates equal spacing of the perturbation parameter ϵ, where ϵ_0 gives the value when the classical Regge pole appears.

ϵ because the classical pole has a branch point at $\epsilon = \epsilon_0$. As a consequence, the quantum pole no longer has convergent expansion in the power series of ϵ for $\epsilon > \epsilon_0$, which means that ϵ_0 is the radium of convergence of the perturbation series for this pole.

Therefore, the indication that classical long-lived states are formed is that the appropriate complex angular momentum pole undergoes a rapid change, shown in Figure 43.

II. SCATTERING ON NONSPHERICAL POTENTIAL

A. Scattering without Energy Transfer

Atom-molecule interaction is in general a nonspherical potential and neglecting this fact, as we did in the previous discussion of vibrational energy transfer, may lead to erroneous conclusions. Nonspherical potential causes a change of molecular rotation; however, in many cases this mode of energy transfer is negligible, e.g., in the collision of an atom with a large molecule. Therefore, in many cases we can neglect rotation of the molecule, as we did in derivation of Equation 454, and this approximation is less serious than neglecting entirely the nonsphericity of the molecule. There are various levels of approximation when describing scattering on nonspherical potential. The simplest approximation is when the molecule is treated as a rigid object. In such a case the set of Equations 454 is independent of n (vibrational quantum number), from which it also follows that there is no energy transfer between the atom and molecule. We expect, therefore, that no long-lived states are formed, except those which are entirely due to the shape of the potential. On this level of approximation the problem is not of much interest; however, it is worth studying because we will understand some of the problems which are encountered when rotations of molecules are taken into account.

In order to extend the idea of complex angular momentum to this kind of collision, we should analyze the coupling elements of Equation 454 in more detail. They are given by

$$v_{lm,l',m'}(r;\quad \theta_M, \phi_M) = \int d\Omega\ Y_l^{M*}V[\cos\theta\cos\theta_M + \sin\theta\sin\theta_M\cos(\phi - \phi_M)]\ Y_{l'}^{m'} \tag{506}$$

where spherical harmonics are defined as

$$Y_l^m = (2\pi\ N_{l,m})^{-1/2}\ P_l^m(\cos\theta)\ e^{im\phi} \tag{507}$$

There is no unique way of making l and m continuous in Equation 506. We will show one way how this can be accomplished if it is assumed that $V(\cos\gamma)$ has expansion of the form

$$V(\cos\gamma) = V_0 + V_1P_1(\cos\gamma) + \ldots \tag{508}$$

The matrix elements of Equation 506 have obvious parametrization

$$v_{l,m,l',m'}(r;\ \theta_M, \phi_M) = e^{i(m'-m)\phi_M}\ \omega_{l,m,l',m'}(r;\ \theta_M) \tag{509}$$

so that the set of equations which we have to solve is

$$\frac{d^2}{dr^2}\psi_{l,m} - \frac{(l+1)\,l}{r^2}\psi_{l,m} - \sum_{l',m'}\omega_{l,m,l',m'}(r;\ \theta_M)\,\psi_{l',m'} = -k^2\,\psi_{l,m} \tag{510}$$

and the scattering amplitude is

$$f(\theta, \phi;\ \theta_M, \phi_M) = \frac{1}{2ik}\sum_{l,m}(2l+1)\,P_l^m[e^{im(\phi-\phi_m)}\,S_{l,m}(\theta_M) - \delta_{m,0}] \tag{511}$$

where the S-matrix is

$$S_{l,m}(\theta_M) = \frac{N_{l,m}^{-1/2}}{2l+1}\sum_{l_0m_0,l'} j^{+}_{l,m,l_0m_0}(j^{-})^{-1}_{l_0m_0,l',0}\,N_{l',0}^{1/2}(2l'+1)\,e^{i\pi(l'+1)} \tag{512}$$

The amplitude in Equation 511 describes scattering of atoms on a molecule which has fixed orientation. Equation 511 is the result of interference of all atoms which are scattered in the angles θ and ϕ, irrespective of their orbital angular momentum. This amplitude is difficult, if not impossible, to measure because the orientation of the molecule is not known. However, from Equation 511 we can obtain all relevant information for this type of collision. For example, the averaged cross-section is

$$\sigma(\theta, \phi) = \frac{1}{4\pi}\int d\Omega_M\,|f|^2 \tag{513}$$

while the amplitude for scattering into the (j, m) rotational state of the molecule is

$$f_{j,m}(\theta, \phi) = \int Y^{*}_{j,m}(\theta_M, \phi_M)\,f\,d\Omega_M \tag{514}$$

If there is no energy transfer into rotations of the molecule, the amplitude of Equation 514 has no meaning; however, if there is very little transfer (e.g., the momentum of inertia of the molecule is large but not infinite), then Equation 514 can give reasonably good results. We will discuss Equation 514, but only for particular m states, for which the amplitude is

$$f_m = \frac{1}{2ik}\sum_l (2l+1)\,P_l^m(S_{l,m} - \delta_{m,0}) \tag{515}$$

from where we obtain Equation 514 by additional averaging over θ_M. The sum of Equation 515 resembles the elastic scattering amplitude, except that the S-matrix is given by Equation 512. We can transform Equation 515 by the Poisson's sum, as in Equation 245, in which case we formally obtain

$$f_m = \frac{1}{ik}\sum_{l=-\infty}^{\infty} e^{-iL\pi}\int_0^{\infty} d\lambda\,\lambda\,P^m_{\lambda-1/2}(S_{\lambda-1/2,m} - \delta_{m,0})\,e^{2\pi iL\lambda} \tag{516}$$

The sum over l in Equation 515 starts from $l = m$, but we have lowered this limit to $l = 0$ because for integer l the Legendre polynomials P_l^m are zero when $l < m$. For continuous l we can use the analytic continuation of P_l^m

$$P_l^m = (-1)^m \frac{(1 - x^2)^{m/2}}{2^m\, m!} \frac{\Gamma(l + m + 1)}{\Gamma(l - m + 1)} F\left(m - l,\; m + l + 1;\; m + 1;\; \frac{1 - x}{2}\right) \tag{517}$$

where F(a, b; c; z) are hypergeometrical functions.[55,57]

In Equation 516 we have formally defined $S_{l,m}$ for continuous l; however, it is not clear how to do it practically. One way of doing this is by defining η with the property

$$\lambda = \frac{1}{2} + \eta + \mathrm{Int}(\lambda) \tag{518}$$

where $-\frac{1}{2} \leqslant \mathrm{Re}(\eta) < \frac{1}{2}$. If we designate $l = \mathrm{Int}(\lambda)$ then the set of Equations 510 is

$$\psi''_{l,m} = \left(\frac{\lambda^2 - 1/4}{r^2} - k^2\right) \psi_{l,m} + \sum_{l',m'} \omega^{(\eta)}_{l,m,l',m'}\, \psi_{l',m'} \tag{519}$$

while the S-matrix is

$$S_{\lambda,m} = \frac{i}{\lambda} N^{-1/2}_{\lambda,m} \sum_{n_0 m_0, l'} j^{+}_{l,m,n_0 m_0}\, j^{-\;-1}_{n_0 m_0,\; l',0}\, \lambda'\, e^{i\pi\lambda'}\, N^{1/2}_{\lambda',0} \tag{520}$$

where $\lambda' = l' + {}^1/_2 + \eta$. If we define

$$\varphi_{l,m} = \frac{1}{2l + 1} N^{-1/2}_{l,m}\, \psi_{l,m} \tag{521}$$

then Equation 520 simplifies and becomes

$$S_{\lambda,m} = i \sum_{n_0 m_0, l'} j^{+}_{l,m,n_0 m_0} (j^{-})^{-1}_{n_0,m_0;l'0}\, e^{i\pi\lambda'} \tag{522}$$

while the set of Equations 519 for 521 is

$$\varphi''_{l,m} = \left(\frac{\lambda^2 - 1/4}{r^2} - k^2\right) \varphi_{l,m} + \lambda^{-1} N^{-1}_{l,m} \sum_{l',m'} \omega^{(\eta)}_{l,m;l',m'}\, \lambda'\, \varphi_{l',m'} \tag{523}$$

where now the matrix elements $\omega_{l,m,l',m'}$, for integer l and l′, are given by

$$\omega_{l,m,l',m'} = \frac{1}{2\pi} \int_{-1}^{1} P_l^m P_{l'}^{m'}\, dx \int_0^{2\pi} d\phi\, e^{i(m' - m)\phi}\, V \tag{524}$$

We can make l and l′ in Equation 524 continuous in two different ways. One way is to evaluate Equation 524 for integer l and l′, and then replace these variables by continuous values. The other way is to replace P_l^m and $P_{l'}^{m'}$ by Legendre functions. In the first approach one must know the analytic expression of the integrals in Equation 524 in order to be able to make l and l′ continuous. This can be done by the use of the expansion in Equation 508 in which case Equation 524 is for the V_L coefficient

$$\omega_{l,m,l',m'}(L) = (-1)^m \left(\frac{N_{l,m} N_{l',m'}(2l+1)(2l'+1)\, 4\pi}{(2L+1)}\right)^{1/2} \begin{pmatrix} l & L & l' \\ 0 & 0 & 0 \end{pmatrix}$$
$$\sum_{M=-L}^{L} (-1)^M \begin{pmatrix} l & L & l' \\ -m & -M & m' \end{pmatrix} Y_L^M(\theta_M, \phi_M) \quad (525)$$

where $\begin{pmatrix} a & b & c \\ d & e & f \end{pmatrix}$ are the "3j" coefficients.* By a suitable choice of the representation of the "3j" coefficients one can replace l and l′ by continuous values.

On the other hand, an advantage of the second approach is that we do not need to know the expansion coefficients in Equation 508. We can obtain Equation 524 by direct numerical integration over the angle θ. However, generalization of P_l^m for continuous values of l is not straightforward. The Legendre functions are singular for $x = -1$, of the form[55]

$$P_l^m(x) \sim (1 + x)^{-m/2} \quad (526)$$

and therefore the integrals in Equation 508 cannot be calculated. However, if in the definition of the spherical harmonics in Equation 507 instead of using P_l^m we use P_l^{-m}, then the matrix elements in Equation 524 would have the form

$$\omega_{l,m,l'm'} = \int_{-1}^{1} P_l^{-|m|} P_{l'}^{-|m'|} V_{m,m'} \quad (527)$$

so that their possible generalization for continuous l and l′ is

$$\omega^{(\eta)}_{l,m,l',m'} = \int_{-1}^{1} P_{l+\eta}^{-|m|-\eta} P_{l'+\eta}^{-|m'|-\eta} V_{m,m'} \quad (528)$$

where $l \geq |m|$ and $l' \geq |m'|$. In such a case the Legendre functions are polynomials of the order $l - |m|$ and $l' - |m'|$, respectively. We can now obtain the elements of Equation 524 without, in fact, calculating the expansion coefficients in Equation 508.

In general, Equations 525 and 528 do not give the same result for continuous l. However, in both cases, as one can show, the set of Equations 523 is singular when $\lambda \to 0$. This type of singularity has been noticed before[33,66,67] in the case of collisions of particles with spin, and it was shown to be an unimportant singularity because it can be removed by simple transformation. Appearance of singularity is the consequence of our choice of the Legendre polynomials as the expansion set for the angular part of the wave function. The choice of the Legendre polynomials was primarily motivated by the fact that they are eigenfunctions of the total angular momentum operator, which has advantages when the potential is spherically symmetric. For nonspherical potentials this advantage is sometimes lost because there is a difference between positive and negative impact parameters. As will be shown in the case of a two-dimensional problem, the scattering at positive and negative impact parameters gives different results, and therefore it is natural to define angular momentum which explicitly distinguishes these two possibilities. The choice of eigenfunctions Y_l^m is not always the most convenient because l has only positive values, and therefore it does not distinguish between positive and negative impact parameters. As the result of this we obtain a nonphysical set

* General theory is found in any book on angular momentum in quantum mechanics, e.g., Rose, M. E., *Elementary Theory of Angular Momentum*, John Wiley & Sons, New York, 1957.

of equations when $\lambda \sim 0$. Nevertheless, the S-matrix in Equation 522 is finite in the limit $\lambda \rightarrow 0$ so that we can use Equation 523 outside the vicinity of that point. The other possibility is to use a different expansion set for the angular part of the wave function. Obvious choices are the functions

$$Y_l^m = e^{il\theta + im\phi} \tag{529}$$

which are not eigenfunctions of L^2. In fact, it is quite straightforward to show that the whole scattering theory can be formulated in terms of these functions, including the elastic problems. For example, the incident plane wave would have the expansion[57]

$$e^{ikr\cos\theta} = \sum_{l=-\infty}^{\infty} e^{i\pi/2 \cdot l} J_l(k\,r)\, e^{il\theta} \tag{530}$$

However, in the scattering problems with spherically symmetric potentials the choice of Equation 529 would lead to a set of coupled equations in l. For spherical potentials this may be an advantage, but when nonspherical potentials are considered, the problem is irrelevant because in any case we are dealing with coupled sets of equations.

B. Two-Dimensional Scattering

We will demonstrate on a two-dimensional model some of the essential features of the scattering problem on nonspherical potentials. We will also assume that the target can rotate, unlike the case in the previous section. This model is analogous to the collinear collision model for the vibrational energy transfer, which means that we use it primarily to understand processes of rotational energy transfer, with the hope that we can describe some of the effects which occur in such collisions. The mathematical description of this collision is greatly simplified in comparison with the three-dimensional case.

The two-dimensional model can give some relevant information about the rotational energy transfer.[68] In particular, it can describe quite well all the effects which are connected with large energy transfer. However, when long-lived states are concerned the model may be less useful because during the time when the atom and molecule are together the out-of-plane rotations can be initiated.

Nevertheless, because of its simplicity it is very instructive to review it and learn some of the problems which are involved in rotational energy transfer.

For the scattering in the plane the Schrödinger equation is

$$\frac{\partial^2 \psi}{\partial x^2} + \frac{\partial^2 \psi}{\partial y^2} + \frac{\mu}{I}\frac{\partial^2 \psi}{\partial \phi^2} = [V(\theta - \phi) - k^2)\,\psi \tag{531}$$

where μ is the reduced mass of the system and I is the momentum of inertia of the molecule. The angle ϕ is the orientation of the molecule with respect to the x-axis, while θ is the polar angle of the atom. When we write ψ as the expansion

$$\psi = \frac{1}{\sqrt{r}} \sum_{J,m} e^{iJ\theta + im\alpha}\, \varphi_m^J(r); \quad J, m = 0, \pm 1, \pm 2, \ldots \tag{532}$$

where $\alpha = \theta - \phi$, then the set of equations for φ_m^J is

$$\varphi_m'' = \left[\frac{1}{r^2}\left\{(J - m)^2 - \frac{1}{4}\right\} - k^2 + \eta\, m^2\right] \varphi_m + \sum_{m'} V_{m,m'}\, \varphi_{m'} \tag{533}$$

where $\eta = \mu/I$. The index J of φ_m we have omitted for convenience. The matrix elements $V_{m,m'}$ are independent of J, and they are given by

$$V_{m,m'} = \frac{1}{2\pi}\int_0^{2\pi} d\alpha \; e^{i\alpha(m-m')} \, V(\alpha) \tag{534}$$

If we assume that $V(\alpha)$ is a function of cos α, then from Equations 533 and 534 we deduce the symmetry property of φ^J_m

$$\varphi^{-J}_{-m} = \varphi^J_m \tag{535}$$

while the S-matrix, which is given by[70]

$$S^J_{m,m'} = \sqrt{\frac{k_m}{k_{m'}}}\,[j^+(j^-)^{-1}]^J_{m,m'} \exp\left[i\frac{\pi}{2}(2J - m - m' + 1)\right] \tag{536}$$

has the property

$$S^{-J}_{m,m'} = e^{-2i\pi J - i\pi(m+m')} \, S^J_{-m,-m'} \tag{537}$$

The scattering amplitude in two dimensions we will define as

$$f_{m,m'}(\theta) = \sum_{J=-\infty}^{\infty} e^{iJ\theta}(S^J_{m,m'} - \delta_{m,m'}) \tag{538}$$

and if we apply the Poisson's summation formula we obtain

$$f_{m,m'}(\theta) = \sum_{l=-\infty}^{\infty} \int_{-\infty}^{\infty} dJ \; e^{iJ\theta}(S^J_{m,m'} - \delta_{m,m'}) \, e^{2il\pi J} \tag{539}$$

where the scattering angle θ is defined in the interval $-180° < \theta < 180°$. The variable J plays now the role of the impact parameter, which is now defined for positive and negative values. Continuation of Equation 539 to the complex J-plane is not straightforward because we must distinguish between positive and negative scattering space, i.e., between $\theta > 0$ and $\theta < 0$, respectively. If $\theta > 0$ then for positive l the integrals in Equations 539 can be continued to the upper half of the J-plane, and if we use Equation 537 in the sum over negative l, the transformed scattering amplitude is

$$\begin{aligned} f_{m,m'}(\theta) = & -\pi \sum_n \frac{\beta^{(n)}_{m,m'}}{\sin(\pi J_n)} \exp[iJ_n(\theta - \pi)] \\ & - \pi \sum_n \frac{\beta^{(n)}_{-m,-m'}}{\sin(\pi J_n)} \exp[-iJ_n(\theta - \pi) - i\pi(m + m')] \\ & + \int_{-\infty}^{\infty} dJ \; S^J_{m,m'} \exp[iJ(\theta - 2\pi)]; \quad \theta > 0 \end{aligned} \tag{540}$$

where J_n are poles of the S-matrix in Equation 536 in the upper half of the J-plane. Similarly, we obtain the scattering amplitude for negative θ, which is related to Equation 540 by

$$f_{m,m'}(-\theta) = f_{-m,-m'}(\theta)\exp[-i\pi(m + m')] \tag{541}$$

The residues $\beta^{(n)}_{-m,-m'}$ in Equation 540 are related to the residues $\beta^{(-n)}_{m,m'}$ for $-J_n$, which means that the second sum in Equation 540 represents waves traveling in the opposite direction to the waves represented by the first sum.

The set of equations from which the poles are calculated resembles very much the set of equations for vibrational energy transfer. The exception is that the centrifugal term contains m, the rotational quantum number. If the off-diagonal elements of $V_{m,m'}$ are assumed to be perturbation, then the unperturbed poles are solutions of the equation

$$\varphi''_m = \left(V_{0,0} + \frac{l_m^2}{r^2} - k^2 + \eta m^2\right)\varphi_n \tag{542}$$

where

$$l_n = J_n^{(m)} - m = J_n^{(-m)} + m \tag{543}$$

This means that the unperturbed poles $J_n^{(m)}$ and $J_n^{(-m)}$ are related and that these poles are not confined to only the first and third quadrants of the J-plane. If we recall, this was the case when we considered inelastic collisions with vibrational energy transfer.

Finding correction to the unperturbed poles, using perturbation theory, is not straightforward because the coupling elements in Equation 534 can be very large. They are large because of the nonspherical character of the repulsive core of the potential, the point which was discussed in the chapter on semiclassical theory. However, very often we can use decoupling schemes which directly take into account the shape of the repulsive core and in this way avoid difficulties.

The theory is based on neglecting the off-diagonal elements of Equation 534. There are two ways how this can be done. In the first approach we transform Equation 531 into cylindrical coordinates and by transformation of angle variables to

$$\alpha = \theta - \phi; \quad \beta = \theta + \frac{1}{\eta\, r^2}\phi \tag{544}$$

we obtain the equation

$$\frac{\partial^2 \varphi}{\partial r^2} + \left(\eta + \frac{1}{r^2}\right)\frac{\partial^2 \varphi}{\partial \alpha^2} + \frac{1}{r^2\eta}\left(\eta + \frac{1}{r^2}\right)\frac{\partial^2 \varphi}{\partial \beta^2} = \left[V(\alpha;\ r) - \frac{1}{4r^2} - k^2\right]\varphi \tag{545}$$

where $\varphi = r^{-1/2}\varphi$. We can now define angular functions as the solution of the equations

$$\left(\eta + \frac{1}{r^2}\right)\frac{d^2 Y}{d\alpha^2} - V(\alpha;\ r)\,Y = \lambda\,Y \qquad \frac{d^2 Z}{d\beta^2} = -\rho^2 Z \tag{546}$$

Since r enters parametrically into Equations 546, the angular functions Y and Z are r-dependent. Therefore, when we make expansion of φ into Y and Z, the coupling elements for the radial wave function will be entirely due to the r-derivatives of these functions. In these elements there will be no problem with hard core, which will be taken care of by the functions Y. These functions can be sometimes obtained analytically, e.g., in the case of ion-dipole interaction.[71] In such a case

$$V(\alpha;\ r) \sim \frac{\cos\alpha}{r^2} + V_0(r) \qquad (547)$$

so that the equation for Y represents the Mathieu equation,* for which solutions are known. In practice one can start from Equation 534 and diagonalize the right side of the equation at each r. The resulting equations are analogous to the adiabatic set of equations, discussed in the previous sections.

Another approach to decoupling schemes is to transform Equation 541 into curvilinear orthogonal coordinates, in which the equipotential line E = V will be identical with one of the coordinates. For example, if the equipotential line is an ellipse, then on this line the radial elliptical coordinate is constant. We will sketch how such a transformation is done for the case of elliptical molecules. We must first transform the kinetic energy of Equation 541 in the rotated coordinate system in which $\phi = 0$. This is done by the transformation

$$\begin{aligned} x' &= x\cos\phi + y\sin\phi \\ y' &= -x\sin\phi + y\cos\phi \end{aligned} \qquad (548)$$

The kinetic energy is now

$$T = \frac{\partial^2}{\partial x'^2} + \frac{\partial^2}{\partial y'^2} + \eta\left(y'\frac{\partial}{\partial x'} - x'\frac{\partial}{\partial y'} + \frac{\partial}{\partial\phi}\right)^2 \qquad (549)$$

In these coordinates we can now define elliptical coordinates by transformation*

$$\begin{aligned} x' &= \frac{a}{2}\,\mathrm{ch}\,\rho\cos\theta \\ y' &= \frac{a}{2}\,\mathrm{sh}\,\rho\sin\theta \end{aligned} \qquad (550)$$

in which case the translational kinetic energy is

$$T_{tr} = \frac{4}{a^2(\mathrm{ch}^2\rho - \cos^2\theta)}\left(\frac{\partial^2}{\partial\rho^2} + \frac{\partial^2}{\partial\theta^2}\right) \qquad (551)$$

while

$$l' = y'\frac{\partial}{\partial x'} - x'\frac{\partial}{\partial y'} = -\frac{1}{\mathrm{ch}^2\rho - \cos^2\theta}\left(\sin\theta\cos\theta\frac{\partial}{\partial\rho} + \mathrm{sh}\,\rho\,\mathrm{ch}\,\rho\frac{\partial}{\partial\theta}\right) \qquad (552)$$

T_{tr} and l' commute; however, l' does not in general commute with potential, and therefore in these coordinates there is no easy way of defining total angular momentum. Furthermore, the Schrödinger equation, which is now given by

$$\frac{8}{a^2(\mathrm{ch}\,2\rho - \cos 2\theta)}\left(\frac{\partial^2\psi}{\partial\rho^2} + \frac{\partial^2\psi}{\partial\theta^2}\right) + \eta\left(l' + \frac{\partial}{\partial\phi}\right)^2\psi = [V(\rho,\theta) - k^2]\,\psi \qquad (553)$$

* General discussions are found in Reference 26.

is more complicated to solve, but the advantage is that the matrix elements

$$V_{m,n} = \frac{1}{2\pi} \int_0^{2\pi} d\theta \, e^{i(m-n)\theta} V(\rho, \theta) \tag{554}$$

do not suffer from the inadequacies of Equation 534. In fact, $V(\rho, \theta)$ can be θ-independent and at the same time be spherically nonsymmetric, i.e., we would be able to describe the rotational energy transfer.

Both Equations 545 and 553 are much better suited for perturbation expansion of poles than Equation 531. However, perturbation theory is only adequate for description of collisions with relatively small energy transfer. When collisions with large energy transfer are considered then one must solve the Schrödinger equation numerically. Even then, Equations 545 and 553 may be easier to solve than Equation 531, but the latter involves less analytical work.

In most cases, energy transfer is almost entirely determined by two kinematic and one dynamic parameter. The kinematic parameters are the momentum of inertia of the molecule and the reduced mass of the system, while nonsphericity of the repulsive core is the dynamic parameter. For example, if a molecule is a nonrotating hard ellipsoid, then the maximal transfer of rotational energy is[68]

$$E_{rot} = \frac{2\,\Delta^2}{I(1 + \eta\Delta^2)^2} \tag{556}$$

where Δ is the difference between two axes of the ellipsoid. It is obvious that if long-lived states are expected to be formed, the value of Equation 556 must be large in addition to the fact that the potential must have a well. However, one should be careful when making such conclusions, because if the minimum of potential is within reach of the large axis of the ellipsoid, the long-lived states are easily destroyed by the multiple collisions between the molecule and atom. Therefore, it is expected that formation of long-lived states in collisions with rotational energy transfer does not often happen, and if it happens then their lifetime is not long. A somewhat better chance of formation of long-lived states occurs when the minimum of potential is outside the reach of the large axis of the molecule. Even then, it is expected that their lifetime is relatively short because of the very efficient mechanism of rotational energy transfer.

C. Rotational Energy Transfer

The theory of rotational energy transfer in atom-molecule collisions is a generalization of the theory of spinless-spinning particle collision, where the potential plays the role of spin-orbit coupling.[72,73] An additional feature of atom-molecule collisions is that each "spin" of a molecule has different energy. For this type of collision it is customary to define states of particular total angular momentum, the feature which was missing in the treatment in Section II.A. If the rotational state of the molecule is described by the function Y_j^m and the state of orbital angular momentum described by Y_l^m, then the state of total angular momentum J is given by

$$Y_J^M = \sum_{m,\mu} \langle l\; m\; j\; \mu | JM\rangle \, Y_l^m \, Y_j^\mu \tag{557}$$

where $\langle l\, mj\mu|JM\rangle$ are the Clebsch-Gordon coefficients. Multichannel equations for radial motion are in this representation

$$\psi''_{l,j} = \left[\frac{l(l+1)}{r^2} - k_j^2\right]\psi_{l,j} + \sum_{l',j'} \langle lj|V|l'\,j'\rangle\,\psi_{l',j'} \qquad (558)$$

where the coupling elements are

$$\langle l\,j|V|l'\,j'\rangle = \sum_{\substack{m,\mu\\ m',\mu'}} \langle l\,m\,j\,\mu|J - M\rangle\langle l'm'j'\mu'|JM\rangle \int d\Omega\, d\Omega_M\, Y_l^m\, Y_j^{\mu}V\, Y_{l'}^{m'}Y_{j'}^{\mu'} \qquad (559)$$

Scattering amplitude is most conveniently represented in the helicity representation, in which the projection axis of the "spin" of the molecule is defined along the velocity vector of the molecule. Following the notation of Calogero et al.[74] we have

$$f_{j,\mu\to j'\mu'}(\theta,\phi) = \frac{\exp[i(\mu' - \mu)\,\phi]}{2i(k_j\,k_{j'})^{1/2}} \sum_J (2J+1)\, d^J_{\mu,\mu'}(\theta)\, F^J_{\mu,\mu'} \qquad (560)$$

where μ and μ' are helicities. The coefficients $A^J_{\mu,\mu'}$ are defined as

$$F^J_{\mu,\mu'} = \sum_{l,l'} \langle l\,0\,j\mu|J\mu\rangle\,\langle l'0\,j'\mu'|\,J\mu'\rangle\,\frac{[(2l+1)(2l'+1)]^{1/2}}{2J+1}$$

$$\cdot\,(S^J_{lj,l'j'} - \delta_{ll'}\,\delta_{jj'}) \qquad (561)$$

where $S^J_{lj,l'j'}$ are the S-matrix elements. The functions $d^J_{\mu,\mu'}(\theta)$ are elements of the rotation matrix which can be written in the form[74]

$$d^J_{\mu,\mu'} = (-1)^{\mu_0-\mu}\left[\frac{(J+\mu_1)!(J-\mu_1)!}{(J+\mu_0)!(J-\mu_0)!}\right]^{1/2} \cdot \left(\frac{1-Z}{2}\right)^{|\mu-\mu'|/2}$$

$$\cdot \left(\frac{1+Z}{2}\right)^{(\mu+\mu')/2} P^{(|\mu-\mu'|,\mu+\mu')}_{J-\mu_1}(Z) \qquad (562)$$

where

$$\mu_1 = \frac{1}{2}(\mu + \mu' + |\mu - \mu'|)$$

$$\mu_0 = \frac{1}{2}(\mu + \mu' - |\mu - \mu'|) \qquad (563)$$

and

$$P_n^{(a,b)}(z) = \binom{n+a}{a} F[-n,\, n+a+b+1;\;\; a+1;\;\; 1/2(1-z)] \qquad (564)$$

We have designated $Z = \cos\theta$. The sum in Equation 560 is very similar to the partial wave decomposition of scattering amplitude with spherically symmetric potentials; however, Equation 562 replaces the Legendre polynomials. Therefore, we could formally apply the Poisson's summation formulas to Equation 560, but in Equation 561 we notice that a problem appears analogous to the one discussed in Section II.A. The problem is how to treat l when J is continuous. One way of solving it is to define the integer variable t by $l = J + t$, where t takes values from the interval $-j \leq t \leq j$. In such a case the sums over l and l' are replaced by the sums over t and t′, respectively, and they are independent of J. Clebsch-Gordon coefficients are zero for all values of t when the triangular relationship is not fulfilled, which ensures that the sums over t and t do not contain more terms than the original sums. We can now formally replace J by the continuous variable; however, in doing so we encounter several difficulties. They were discussed in Section II.A, but here there is a complication because of the additional degree of freedom, i.e., the rotations of the molecule.

In order to make J continuous in Equation 558 we will assume that the matrix elements in Equation 559 are calculated for integer J and then this variable is made continuous (see discussion in Section II.A on various ways of making angular momentum a continuous variable). This requires analytic continuation of the Clebsch-Gordon coefficients for continuous J. According to the suggestion of Charap and Squires[72,73] this is possible from the representation

$$\langle J + t, m\, j\, \mu | JM\rangle = N^J_{tj} \sum_\nu \frac{(-1)^{\nu+j+\mu}}{\nu!} \frac{(J + j + m - \nu)!(J + t - m + \nu)!}{(-t + j - \nu)!(J + m + \mu - \nu)!(J + t - j - m - \mu + \nu)!} \quad (565)$$

where

$$N^J_{tj} = \left[(2J + 1)\frac{(2J + t - j)!(-t + j)!(t + j)!(J + m + \mu)!(J - m - \mu)!}{(2J + t + j + 1)!(J + t - m)!(J + t + m)!(j - \mu)!(j + \mu)!}\right]^{1/2} \quad (566)$$

We notice that for continuous J the normalization coefficients have square root branch points, which are therefore also present in Equation 559. In addition to these branch points the matrix elements of Equation 559 of potential have poles similar in nature to the poles discussed in Section II.A. It was shown by Charap and Squires that branch points and poles can be removed by the following transformation of the wave function:

$$\varphi^J = A^J \psi^J \quad (567)$$

The potential matrix elements in Equation 559 transform by

$$V^J = A^J V^J (A^J)^{-1} \quad (568)$$

where the matrix elements of A^J are given by

$$A^J_{tj,l'j'} = \delta_{j,j'}\left[\frac{(2l' + 1)(J + t)!(J - t)!}{2J + 1}\right]^{1/2} \langle l'0\, j't|Jt\rangle \quad (569)$$

and its inverse

$$(A^J)^{-1}_{lj,l'j'} = \delta_{j,j'}\left[\frac{(2l + 1)}{(2J + 1)(J + t')!(J - t')!}\right]^{1/2} \langle l0jt'|Jt'\rangle \quad (570)$$

The matrix elements of V^J are independent of J, and in order to show this we write V as the expansion

$$V = \sum_L T^{(0)}_L V_L(r) \quad (571)$$

where $T_L^{(0)}$ is the scalar tensor operator which has the form

$$T_L^{(0)} = \sum_{\nu} \langle L\nu L - \nu|00\rangle\, Y_L^{\nu}(l)\, Y_L^{-\nu}(j) \tag{572}$$

The orbital and rotational angular tensors are $Y_L^n(l)$ and $Y_L^{-\nu}$ (j), respectively. It can be shown that the matrix elements in Equation 559 are

$$V^J_{lj,l'j'} = \sum_L V_L(r) \begin{Bmatrix} l & j & J \\ j' & l' & L \end{Bmatrix} \frac{\langle l\|Y_L\|l'\rangle\langle j\|Y_L\|j'\rangle}{\sqrt{2L+1}} (-1)^{L+J+l+j'} \tag{573}$$

where

$$\langle a\|Y_b\|c\rangle = \left[\frac{(2c+1)(2b+1)}{4\pi}\right]^{1/2} \langle b\,0\,c\,0|a\,0\rangle \tag{574}$$

and $\{\ldots\}$ are the "6j" coefficients. The transformed matrix elements in Equation 568 are now

$$V^J_{jt,j't'} = \frac{\delta_{t,t'}}{4\pi} \sum_L V_L(r)\langle L\,0\,j\,t|j't\rangle\, \langle L\,0\,j'0|j\,0\rangle\sqrt{2L+1} \tag{575}$$

which means that they are independent of J and, therefore, free from singularities. These matrix elements resemble the matrix elements in Equation 534 in the two-dimensional model, except that angular functions are here spherical harmonics rather than trigonometric functions.

Transformation of the centrifugal term is obtained from Equation 568 if potential is replaced by $l(l + 1)$. If we use the recurrence relationship for the Clebsch-Gordon coefficients we obtain

$$\begin{aligned} l(l+1)\,\langle l0j\,t|Jt\rangle &= [(j+t)(j-t+1) + (J-t)(J+t+1)]\,\langle l0j\,t|\,JT\rangle \\ &\quad -[(j+t)(j-t+1)(J+t)(J-t+1)]^{1/2}\,\langle l0j\,t-1\,|\,Jt-1\rangle \\ &\quad -[(j-t)(j+t+1)(J-t)(J+t+1)]^{1/2}\,\langle l0j\,t+1\,|Jt+1\rangle \end{aligned} \tag{576}$$

which gives for the matrix elements in Equation 568

$$\begin{aligned} [A^J\, l(l+1)\,(A^J)^{-1}]_{jt,jt'} &= \delta_{j,j'}\delta_{t,t'}[(j+t)(j-t+1) + (J-t)(J+t+1)] \\ &\quad + \delta_{j,j'}\delta_{t,t'-1}(J-t)\,[(j-t)(j+t+1)]^{1/2} + \delta_{j,j'}\,\delta_{t,t'+1}(J+t) \\ &\quad \cdot [(j+t)(j-t+1)]^{1/2} \end{aligned} \tag{577}$$

being free from singularities, including the branch points. Therefore, it was shown that by transformation of Equation 567 we have obtained a set of equations for φ^J which are free from all spurious singularities. These equations can now be generalized for continuous J. It should be noted that the channel energies k_j^2 in Equation 558 are invariant to transformation of Equation 568.

Going back to the scattering amplitude in Equation 560 we notice that the sums in Equation 561 represent transformation of the S-matrix, of the form in Equation 568. Therefore, we can write

$$F^J_{\mu,\mu'} = \left[\frac{(J + \mu')!(J - \mu')!}{(J + \mu)!(J - \mu)!}\right]^{1/2} (S^J_{J\mu,J'\mu'} - \delta_{JJ'}\, \delta_{\mu,\mu'}) \quad (578)$$

where the transformed S-matrix is obtained either by transformation of Equation 568 or directly from the solutions φ^J. The second case is sometime more convenient, because we directly solve the set of equations for φ^J. For large r the radial wave functions φ^J behave as

$$\varphi^J \sim e^{ikr}\, x^+ + e^{-ikr}\, x^- \quad (579)$$

where $x^\pm$ are the Jost's functions so that the S-matrix is given by

$$S^J_{J\mu,J'\mu'} = -\sqrt{\frac{k_J}{k_{J'}}}\, e^{i\frac{\pi}{2}(\mu + \mu' + 2J)}\, [x^+(x^-)^{-1}]_{J\mu,J'\mu'} \quad (580)$$

The scattering amplitude is now

$$f_{J\mu\rightarrow J'\mu'}(\theta, \phi) = \frac{\exp[i(\mu' - \mu)\, \phi]}{2i(k_J\, k_{J'})^{1/2}} \sum_J (2J + 1)\, \mathbf{d}^J_{\mu\mu'}(\theta)$$

$$\cdot\, (S^J_{J\mu,J'\mu'} - \delta_{JJ'}\, \delta_{\mu\mu'}) \quad (581)$$

where

$$\mathbf{d}^J_{\mu\mu'} = d^J_{\mu\mu'}\left[\frac{(J + \mu')!(J - \mu')!}{(J + \mu)!(J - \mu)!}\right]^{1/2} \quad (582)$$

In this form, it resembles the scattering amplitude for collisions with spherically symmetric potentials, where $d^J{}_{\mu\mu'}$ plays the role of the Legendre polynomials. In fact, it can be shown that

$$d^{\mu_1}_{-\mu-\mu'}\, d^J_{\mu\mu'} = d^{\mu_1}_{\mu'\mu}\, d^J_{\mu\mu'} = \sum_{l=-\mu_1}^{\mu_1} P_{J+l}(\cos\theta)\langle J\mu\, \mu_1 - \mu|J + l\, 0\rangle$$

$$\langle J\mu'\mu_1 - \mu'|J + l\, 0\rangle \quad (583)$$

where μ_1 is defined by Equation 563, which gives explicit relationship between scattering amplitude and Legendre polynomials.

The sum in Equation 581 begins with $J = \mu_1$, but for the use of the Poisson's summation formulas it would be convenient to extend its limit to $J = 0$. Since $d^J_{\mu,\mu'}$ is zero when $\mu_0 \leq J < \mu_1$, then without difficulties we can replace μ_1 in Equation 581 by μ_0, if the latter is greater than or equal to zero. If $\mu_0 < 0$, then the lower limit of J is zero. Therefore, we can write for Equation 581

$$f_{J\mu\rightarrow J'\mu'} = \sum_{J=0} (2J + 1)\, \mathbf{d}^J_{\mu,\mu'}(\theta)\, (S^J_{J\mu,J'\mu'} - \delta_{JJ'}\, \delta_{\mu,\mu'})$$

$$- \sum_{J=0}^{\mu_0 - 1} (2J + 1)\, \mathbf{d}^J_{\mu,\mu'}(\theta)\, (S^J_{J\mu,J'\mu'} - \delta_{JJ'}\, \delta_{\mu\mu'}) \quad (584)$$

where we have omitted nonessential factors. We can now apply the Poisson's summation formulas to the first sum, while the second sum must be treated on its own. However, we should notice some symmetry properties of the reduced rotation matrix and the S-matrix which will be useful in transformation. It can be verified that

$$\mathbf{d}^{J}_{\mu,\mu'} = \mathbf{d}^{-J-1}_{-\mu',-\mu} \tag{585}$$

and

$$\mathbf{S}^{J}_{j\mu,j'\mu'} = \mathbf{S}^{-J-1}_{j-\mu',j'-\mu}\, e^{2i\pi\left(J+\frac{1}{2}\right)} \tag{586}$$

The first sum in Equation 584 is transformed according to Equation 244 which gives

$$f^{(1)}_{j\mu\to j'\mu'} = 2 \sum_{M=-\infty}^{\infty} (-1)^{M} \int_0^{\infty} d\,J\,J\,\mathbf{d}^{J-1/2}_{\mu,\mu'} \cdot \mathbf{S}^{J-1/2}_{j\mu,j'\mu'}\, e^{2\pi i M J} \tag{587}$$

where we have assumed that $j \neq j'$. The sum over the negative M is equal to

$$\sum_{M<0} = \int_0^{-\infty} d\,J\,J\,\mathbf{d}^{J-1/2}_{-\mu',-\mu}\, \mathbf{S}^{J-1/2}_{j-\mu',j'-\mu}\, \frac{e^{-i\pi J}}{\cos(\pi J)} \tag{588}$$

The integral can be formally shifted to the positive imaginary axis, in which case the sum in Equation 588 gets contribution from poles of the S-matrix in the second quadrant. Such poles cannot be excluded as was the case with the spherically symmetric potentials. Therefore, the sum in Equation 588 is

$$\sum_{M<0} = -2\pi i \sum_{n} J_n\, \mathbf{d}^{J_n-1/2}_{-\mu',-\mu}\, \frac{e^{-i\pi J_n}}{\cos(\pi J_n)}\, \beta^{(n)}_{j-\mu',j'-\mu} \tag{589}$$

where we have neglected the integral along the positive imaginary axis. $\beta^{(n)}_{a,b}$ is the residue of the S-matrix at the pole J_n, which in Equation 589 is from the second quadrant of the J-plane.

Similarly we transform the sum over the positive M in Equation 587, so that finally we have

$$f^{(1)}_{j\mu\to j'\mu'} = \int_0^{\infty} d\,J\,J\,\mathbf{d}^{J-1/2}_{\mu,\mu'}\, \mathbf{S}^{J-1/2}_{j\mu,j'\mu'}$$

$$+ 2\pi i \sum_{n} J_n\, \mathbf{d}^{J_n-1/2}_{\mu,\mu'}\, \frac{e^{i\pi J_n}}{\cos(\pi J_n)}\, \beta^{(n)}_{j\mu,j'\mu'} \tag{590}$$

where now the sum over the poles includes those from the first and fourth quadrants of the J-plane. In derivation of Equation 590 we have used Equations 585 and 586.

Therefore, we have obtained a transformed scattering amplitude, which is formally very similar to the scattering amplitude for spherically symmetric potentials. However, in this case, properties of the J-poles are not yet very well understood, and because of this we cannot say much about properties of the long-lived states in such collisions.

We should mention that our analysis was based on the helicity representation of the scattering amplitude. In this representation we obtained the relatively simple form of the scattering amplitude, but this does not necessarily mean that it is the most convenient for description of the collision processes. We have seen that the most efficient energy transfer is in the plane, and in this configuration the molecule can rotate, relative to the incoming atom, in the positive or negative sense. In the two-dimensional model these two senses of rotation ar distinguished according to the sign of j. However, in helicity representation both rotations correspond to $\mu = 0$, which implies that the left and right scattering spaces are not distinguished, and the result of this is that the scattering amplitude in Equation 581 is symmetric apart factor with respect to these two spaces, i.e., to transformations $\theta \rightarrow \theta$ and $\phi \rightarrow \phi + \pi$.

APPENDIX A

We want to investigate the inverse of the integral

$$\varphi_0 = \int d^3 k\, A(k)\, \varphi(\mathbf{k}) \tag{A1}$$

where φ_0 represents the initial wave packet, and $\varphi(\mathbf{k})$ are solutions of the equation

$$\Delta\, \varphi_k = (V - k^2)\, \varphi_k \tag{A2}$$

with boundary condition for $r \rightarrow \infty$

$$\varphi_k \sim e^{i\mathbf{kr}} + \frac{e^{ikr}}{r} f(\hat{r}, \hat{k}) \tag{A3}$$

where $\hat{k} = \frac{\mathbf{k}}{k}$ determines the direction of the incident wave and $\hat{r} = \frac{\mathbf{r}}{r}$ determines the scattering angle. It can be easily shown from Equation A2 that φ_k satisfies the equation

$$\nabla\, \mathbf{I}_{\mathbf{k}',\mathbf{k}} = \nabla(\varphi_{\mathbf{k}'} \nabla\, \varphi_{\mathbf{k}} - \varphi_{\mathbf{k}}\, \nabla\, \varphi_{\mathbf{k}'}) = (k'^2 - k^2)\, \varphi_{\mathbf{k}'}\, \varphi_{\mathbf{k}} \tag{A4}$$

or

$$(k'^2 - k^2) \int_V \varphi_{\mathbf{k}'}\, \varphi_{\mathbf{k}}\, dV = \oint_S ds(\varphi_{\mathbf{k}'} \nabla \varphi_{\mathbf{k}} - \varphi_{\mathbf{k}} \nabla \varphi_{\mathbf{k}'}) \tag{A5}$$

where the integral over S encircles the whole volume space V. If we write the expansion

$$e^{i\mathbf{kr}} = 4\pi \sum_{l=0}^{\infty} i^l j_l(Kr) \sum_{m=-l}^{l} Y_l^{*m}(\hat{r})\, Y_l^m(\hat{k}) \tag{A6}$$

where j_ℓ is the spherical Bessel function and Y_ℓ^m are the spherical harmonics, then the radial component of the flux $\mathbf{I}_{\mathbf{k}',\mathbf{k}}$ is for $r \rightarrow \infty$

$$\begin{aligned} \hat{r}\, \mathbf{I}_{\mathbf{k}',\mathbf{k}} &= \frac{8\,\pi^2}{k^2 r^2} (k + k') \sin r(k' - k)\, \delta(\hat{r} - \hat{k})\, \delta(\hat{r} + \hat{k}') \\ &+ \frac{2\,\pi}{r^2} \left[\frac{k + k'}{k} e^{ir(k'-k)} f(\hat{r}, \hat{k}')\, \delta(\hat{r} + \hat{k}) - \frac{k + k'}{k'} e^{-ir(k'-k)} \right. \\ &\left. f(\hat{r}, \hat{k})\, \delta(\hat{r} + \hat{k}') \right] \end{aligned} \tag{A7}$$

In the derivation of Equation A7 we have neglected terms of the order r^{-3} or lower, and we have used the property of completeness of the spherical harmonies

$$\sum_{l,m} Y_l^{m*}(\hat{r})\, Y_l^m(\hat{k}) = \delta(\hat{r} - \hat{k}) \tag{A8}$$

If now Equation A7 is replaced in the integral in Equation A5 we obtain

$$(k'^2 - k^2) \int \varphi_{k'} \varphi_k \, dV = \frac{8\,\pi^2}{k^2} (K + k') \sin r(k' - k)\, \delta(\hat{k} + \hat{k}')$$

$$+ 2\pi \frac{k + k'}{k} [e^{ir(k'-k)} f(-\hat{k}, \hat{k}') - e^{-ir(k'-k)} f(-\hat{k}', \hat{k})] \quad \text{(A9)}$$

For $r = \infty$ we can use the definition of the Dirac's delta function

$$\lim_{r\to\infty} \frac{\sin r\, k}{k} = \pi\, \delta(k) \quad \text{(A10)}$$

so that we finally obtain

$$\int \varphi_k \varphi_{-k'} \, dV = \frac{8\,\pi^3}{k^2} \delta(k' - k)\, \delta(\hat{k} - \hat{k}')$$

$$+ \frac{2\,\pi^2 i}{k} \delta(k' - k)[f(\hat{k}', \hat{k}) + f(-\hat{k}, -\hat{k}')] \quad \text{(A11)}$$

The expression in Equation A11 is valid for a general nonspherical potential; however, for symmetric potentials there is equality

$$f(\hat{k}', \hat{k}) = f(-\hat{k}, -\hat{k}') \quad \text{(A12)}$$

so that Equation A11 simplifies.

To obtain the inverse of Equation A1 we calculate the integral

$$\int \varphi_0 \varphi_{-k'} \, dV = \int d^3 k\, A(\mathbf{k}) \int \varphi_k \varphi_{-k'} \, dV \quad \text{(A13)}$$

From Equation A11, and for sperhical potentials, we have

$$\int \varphi_0 \varphi_{-k'} \, dV = 8\,\pi^3 A(\mathbf{k}') + 4\,\pi^2 ik' \int d\Omega_k A(k', \Omega_k)\, f(\hat{k}', \hat{k}) \quad \text{(A14)}$$

It is essential now to assume that φ_0 is significant only far away from the scattering region, i.e., for $r \rightarrow \infty$. In such a case φ_{-k}, can be replaced by Equation A3 from which it follows immediately that

$$A(\mathbf{k}) = \frac{1}{(2\pi)^3} \int \varphi_0 \, e^{-kr} \, dV \quad \text{(A15)}$$

is the solution of Equation A14.

APPENDIX B

We will show here how to solve the equation

$$\psi'' = (V - k^2)\,\psi = -p^2\,\psi \tag{B1}$$

in the limit $h \rightarrow 0$. If we define a new variable

$$\xi = \int_{r_0}^{r} p\, dr' \tag{B2}$$

and replace ψ by

$$\psi = p^{-1/2}\,\varphi \tag{B3}$$

then the equation for φ is

$$\frac{d^2\varphi}{d\xi^2} = \overset{\circ\circ}{\varphi} = -\varphi - Q\,\varphi \tag{B4}$$

where Q is

$$Q = \frac{1}{4p^4}\left(\frac{5}{4}\frac{V'^2}{p^2} + V''\right) \tag{B5}$$

We notice that Q is of the order h^2, and therefore it can be treated as perturbation in Equation B4, except in the vicinity of $r = r_0$ where $p^2 = 0$. In this vicinity Q has the approximate form

$$Q \sim \frac{Q_{-3}}{(r - r_0)^3} + \frac{Q_{-2}}{(r - r_0)^2} + \ldots \tag{B6}$$

which can be written as an expansion

$$Q \sim \frac{Q_{-3}}{\xi^2} + \frac{Q_{-2}}{\xi^{4/3}} + \frac{Q_{-1}}{\xi^{2/3}} + \ldots \tag{B7}$$

The coefficients are found by defining $x = \xi^{2/3}$ in which case $Q\,\xi^2$ has the expansion

$$Q\,\xi^2 = Qx^3 = Q_{-3} + xQ_{-2} + x^2\,Q_{-1} + \ldots \tag{B8}$$

from which it is straightforward to obtain Q_n. For example, Q_{-3} is

$$\lim_{r\rightarrow r_0} x^3 Q = \lim_{r\rightarrow r_0} \frac{\xi^2}{4p^6}\left(\frac{5}{4}V'^2 + V''p^2\right) = \frac{5}{36} \tag{B9}$$

where we have used the fact that $\xi' = p$. Similarly we obtain

$$Q_{-2} = 0 \tag{B10}$$

while Q_{-1} is not essential for our purpose. Therefore, we can write Equation B4 as

$$\frac{d^2 \varphi}{d \xi^2} = -\left(1 + \frac{5}{36\xi^2}\right) \varphi - \mathbf{Q} \varphi \tag{B11}$$

where now $\mathbf{Q}$ is defined as $Q = Q_{-3}/\xi^2 + \mathbf{Q}$. The zero order solution of Equation B11 is

$$\varphi = \xi^{1/2} Z_{1/3}(\xi) \tag{B12}$$

where $Z_{1/3}(x)$ is one of the Bessel functions of the order $^1/_3$. A general solution of Equation B11 is given in the form of the integral equation

$$\varphi = \xi^{1/2} Z_{1/3}(\xi) + \int_0^{\xi} d\,\xi' [H^{(1)}_{1/3}(\xi)\, H^{(2)}_{1/3}(\xi') - H^{(1)}_{1/3}(\xi')\, H^{(2)}_{1/3}(\xi)] \cdot (\xi\xi')^{1/2}\, \mathbf{Q}\, \varphi \tag{B13}$$

where the turning point $r = r_0$ was taken as the lower limit of integration. It can be shown that for $\xi' \rightarrow 0$ the integrand behaves as $\xi^{-1/3}$, which is an integrable singularity, and, therefore, iterations of Equation B13 produce absolutely convergent series in powers of h. This means that we have obtained the solution of Equation B1 in the limit $h \rightarrow 0$, which is represented by Equation B12.

APPENDIX C

Solution of multichannel Equation 320 is given in the form of the Volterra integral equation

$$\psi = \psi_0 + \frac{\epsilon}{2i} K^{-1} \int_{r_0}^{r} [f_1(r)\, f_2(r') - f_1(r')\, f_2(r)]\, W^1\, \psi\, dr^1 \tag{C1}$$

Iteration of this integral equation produces an absolutely convergent series for all finite r and for all reasonable W^1. In particular we are interested in the form of solution ψ for $r \rightarrow \infty$. A first choice would be, as it is usually accepted,

$$\psi \sim e^{ikr} J^+ + e^{-ikr} J^- \tag{C2}$$

If we put this form in Equation C1 and estimate the integrand near $r' \sim r \rightarrow \infty$, then we obtain

$$\psi \sim e^{ikr} J_0^+ + e^{-ikr} J_0^- + \frac{\epsilon}{2i} K^{-1} \int_{r_0}^{r} [e^{ik(r-r')} - e^{-ik(r-r')}]$$

$$W^1(e^{ikr'} J^+ + e^{-ikr'} J^-)\, dr' \tag{C3}$$

If all k_n belong to open channels, and W^1 goes reasonably fast to zero for $r \rightarrow \infty$, then indeed from Equation C3 we again obtain a linear combination as Equation C2, i.e., the integral in Equation C3 is finite. However, if some channels are closed, and W^1 does not decay fast enough for $r \rightarrow \infty$, e.g., as some exponential function $e^{-\alpha r}$, then the estimate of the integral in Equation C3 is

$$\psi \sim \frac{\epsilon}{2i} K^{-1} \int^{r} [e^{ik(r-r')} - e^{-ik(r-r')}]\, W^1 e^{-ik_c r'} J^-\, dr' \tag{C4}$$

where the index C of K refers to the closed channel wave numbers. The other terms in Equation C3 can be put in the form in Equation C2. The integral in Equation C4 is infinite in the limit $r \rightarrow \infty$ and has a general form

$$\psi \sim U\, e^{-ik_c r} J^- = U\, f_{2c}\, J^- \tag{C5}$$

Therefore, a more general asymptotic behavior of ψ than C2 is

$$\psi \sim e^{ikr} J^+ + e^{-ikr} J^- + U\, e^{-ik_c r} J^- \tag{C6}$$

The Jost function J^- is now obtained by comparing Equations C3 and C6, and this gives

$$J^- = J_0^- - \frac{\epsilon}{2i} K^{-1} \int_{r_0}^{\infty} f_1(r')\, W^1\, \psi_F\, dr' \tag{C7}$$

where the index F of ψ refers to only those components of ψ which give finite value of the integral. For example, if W^1 decays fast enough exponentially or it is a cut-off potential, then all components of ψ are included. Also, we include those components of ψ which increase exponentially, but together with f_1 give the integrand which goes to zero for $r \rightarrow \infty$.

We notice from Equation C6 that the elements of ψ with open channel indexes are dominated by the exponentially increasing functions. However, one should recall that Equation C6 itself does not enter the scattering solution but the form $\psi(J^-)^{-1}$, for which one can easily show that the elements with open channel indexes are all finite and oscillatory, as they should be. Therefore, there is no contradiction between the asymptotic form of Equation C6 and scattering solution.

This again points to the fact that closed channels are artifacts of our way of solving quantum problems. As long as we do not know any other way of solving such problems, but by expanding ψ in basis sets, we will always be working with difficult problems connected with closed channels.

APPENDIX D

Here we will solve the nearly degenerate perturbation problem for Regge poles and residues. For simplicity we will assume that all nearly degenerate poles occur in the closed uncoupled channels. By the nearly p-fold degenerate problem we will understand the case when $j_\mu^{(0)}$, which is defined in Equation 308, is a root or approximate root of p uncoupled Jost functions $(J_0^-)_n$. For simplicity we will assume that these are the first p functions, i.e.,

$$\{J_0[j_\mu^{(0)}]\}_n \approx 0; \quad n = 1, 2, \ldots, p \tag{D1}$$

where from now on we will omit the minus sign of the Jost function, which is implicitly assumed.

The Jost function J^- is represented from Equation 325 as a power series in ϵ, i.e.,

$$J = J_0 + J_1 + \frac{\epsilon^2}{2} J_2 + \ldots \tag{D2}$$

where J_0^- is diagonal. For $l = j_\mu$ we can write the first p function of J_0 in a series

$$[J_0(j_\mu)]_n = [J_0(j_n)]_n + (j_\mu - j_n)(J_0')_n + \frac{(j_\mu - j_n)^2}{2}(J_0'')_n \tag{D3}$$

where j_n is the zero of $(J_0)_n$ which is close to j_μ. If now j_μ is represented as the series in Equation 308, and the difference $j_\mu^{(0)} - j_n$ is formally given the order ϵ, then Equation D3 is

$$[J_0(j_\mu)]_n = [j_\mu^{(0)} - j_n + j_\mu^{(1)}](J_0')_n + \frac{\epsilon^2}{2}\{j_\mu^{(2)}(J_0')_n$$
$$+ [j_\mu^{(0)} + j_\mu^{(1)} - j_n]^2 (J_0'')_n\} \tag{D4}$$

J_1 is

$$(J_1)_{m,n} = -\frac{1}{2ik_m}\int_{r_0}^{\infty} f_{1m}(j_\mu)\, W_{m,n}^1\, \psi_n(j_\mu)\, dr; \quad m, n \leqslant p \tag{D5}$$

where we have explicitly indicated that the wave functions are functions of j_μ. This matrix element can be written as

$$(J_1)_{m,n} = -\frac{1}{2ik_m}\int_{r_0}^{\infty} f_{1m}(j_m)\, W_{m,n}^1\, \psi_n(j_n)\, dr$$
$$- \frac{\epsilon}{2ik_m}[j_\mu^{(0)} + j_\mu^{(1)} - j_m]\int_{r_0}^{\infty} f_{1m}' W_{m,n}^1\, \psi_n\, dr$$
$$- \frac{\epsilon}{2ik_m}[j_\mu^{(0)} + j_\mu^{(1)} - j_n]\int_{r_0}^{\infty} f_{1m}\, W_{m,n}^1\, \psi_n'\, dr \tag{D6}$$

where f' and ψ' designate derivative with respect to l. We can now define a matrix U which diagonalizes that part of the submatrix $(J_0^{-1})_m^{-1} \cdot (J_1^-)_{m,n}$ is of the order ϵ, i.e.,

$$U^{-1}(\{[j_\mu^{(0)} + j_\mu^{(1)}]\, I - j\} + (J_0^1)^{-1}\, J_1^0)\, U = \lambda \tag{D7}$$

where J_1^0 means that we have taken in Equation D6 only the matrix elements of the order ϵ^0. With the matrix U we now make the following transformation of J^-

$$\begin{vmatrix} U^{-1} & 0 \\ 0 & I \end{vmatrix} \begin{vmatrix} (J_0^1)^{-1} & 0 \\ 0 & I \end{vmatrix} J \begin{vmatrix} U & 0 \\ 0 & I \end{vmatrix} = \begin{vmatrix} U^{-1}(J_0')^{-1} J_{11}U, & U^{-1}(J_0')^{-1} J_{12} \\ J_{21} U, & J_{22} \end{vmatrix} \quad \text{(D8)}$$

and as the result we obtain a transformed matrix J^T. Its determinant is of the form

$$\det(J^T) = \epsilon^p \det \left| \begin{array}{cc|cc} \lambda_1 & O(\epsilon) & & \\ O(\epsilon) & \lambda_p & & O(\epsilon) \\ \hline & & J_{p+1} & O(\epsilon) \\ & O(\epsilon^c) & O(\epsilon) & J_N \end{array} \right| \quad \text{(D9)}$$

where we have designated by $O(\epsilon)$ the leading power in ϵ of the appropriate block matrix. $\det(J^T)$ can be developed in the power series in ϵ which is up to the order ϵ^{p+1}

$$\det(J^T) = \epsilon^p(\lambda_1\lambda_2 \ldots \lambda_p J_{p+1} \ldots J_N + \epsilon C) \quad \text{(D10)}$$

where C is some complicated function of the matrix elements of J^T. Since it is required that $\det(J^T) = 0$ for all ϵ it follows that

$$\lambda_1 \lambda_2 \ldots \lambda_p = 0; \quad C = 0 \quad \text{(D11)}$$

which means that one of λ_m is zero, giving

$$j_\mu^{(1)} = -\{U^{-1}[j_\mu^{(0)} I - j] U\}_{m,m} - [U^{-1}(J_0')^{-1} J_1^0 U]_{m,m} \quad \text{(D12)}$$

Therefore, j_μ' has p different values, and the initial degeneracy of the problem is destroyed in the first order of perturbation. Of course, that is when $j_\mu^{(0)} I - j = 0$. However, in the case of near degeneracy one can show that treating the problem as a degenerate one is significantly better then treating the same problem as the nondegenerate case.

If λ_m is zero, then C simplifies and the equation $C = 0$ becomes

$$\begin{aligned} & j_\mu^{(2)} + (U^{-1}(J_0')^{-1}\{[j_\mu^{(0)} + j_\mu^{(1)}] I - j\}^2 J_0'' U)_{m,m} \\ & - i\langle U^{-1}(J_0')^{-1} K^{-1}(\{[j_\mu^{(0)} + j_\mu^{(1)}] I - j\} \int_{r_0}^{\infty} f_1 W^1 \psi_0 \, dr \\ & + \int_{r_0}^{\infty} f_1 W^1 \psi_0' \, dr\{[j_\mu^{(0)} + j_\mu^{(1)}] I - j\}) U\rangle_{m,m} \\ & + \frac{1}{2}\left\{U^{-1}(J_0')^{-1} K^{-1} \int_{r_0}^{\infty} f_1 W^1 K^{-1} \int_{r_0}^{r} [f_1(r) f_2(r') - f_1(r') f_2(r)] W^1 \psi_0 dr' U\right\}_{m,m} \\ & + \frac{1}{2} \sum_{n=p+1}^{N} (J_0)_n^{-1} \left[U^{-1}(J_0')^{-1} K^{-1} \int_{r_0}^{\infty} f_1 W^1 \psi_0\right]_{m,n} \left(K^{-1} \int_{r_0}^{\infty} f_1 W^1 \varphi_0 dr \, U\right)_{n,m} = 0 \quad \text{(D13)} \end{aligned}$$

from which we obtain $j_\mu^{(2)}$. Most of the terms in Equation D13 are real (or imaginary); therefore, they contribute only to the shift of the Regge pole. The imaginary part of $j_\mu^{(2)}$ is obtained from the last two terms and only when the intermediate sum involves the open channel. We have

$$\mathrm{Im}[j_\mu^{(2)}] = -\frac{1}{4}\sum_{n=\mathrm{open}}\frac{1}{J_{on}J_{on}^+}\left[U^{-1}(J_0')^{-1}|K|^{-1}\int_{r_0}^{\infty} f_1 W^1\psi_0 dr\right]_{m,n}$$

$$\left(K^{-1}\int_{r_0}^{\infty}\psi_0\, W^1\,\psi_0\, U\, dr\right)_{n,m} \qquad \text{(D14)}$$

We can use the property for the pole 1

$$i\,k\,j^+\frac{\partial j^-}{\partial l} = -(l + 1/2)\int_{r_0}^{\infty}\frac{\psi^2}{r^2}dr = -(l + 1/2)\,A^2 \qquad \text{(D15)}$$

in order to define a unitary matrix T by $U = A^{-1}(j + {}^1/_2)^{-1/2}T$. The imaginary part of $j^{(2)}$ now becomes simpler

$$\mathrm{Im}[j_\mu^{(2)}] = -\frac{1}{4}\sum_{n=\mathrm{open}}\frac{1}{K_n J_{on} J_{on}^+}\left(\sum_{q=1}\int_{r_0}^{\infty}\psi_{on}W^1_{n,q}\psi_{oq}dr\,\frac{T_{qm}}{A_2\sqrt{j_q + 1/2}}\right)^2 \qquad \text{(D16)}$$

The coefficients $Q_n^{(j)}$ in Equation 314 can be obtained by using Equation D10. The function $D(-k_n)$ is therefore

$$D(-k_n) = \epsilon^{p+1}\,C(-k_n) \qquad \text{(D17)}$$

where it was assumed that $\lambda_m = 0$, because the change $k_n \rightarrow -k_n$ does not affect either λ in Equation D10. However, this change affects C, and if we take into account Equation D13 we obtain $C(-k_n)$

$$C(-k_n) = -(\lambda_1\lambda_2\ldots\lambda_p)^{(m)}\,(J_{op+1}\,J_{op+2}\cdots J_{ON})$$

$$\frac{1}{4k_n\,J_{on}^2}\left(\sum_{q=1}^{p}\int_{r_0}^{\infty}\psi_{on}W^1_{nq}\psi_{oq}\,dr\,\frac{T_{qn}}{A_0\sqrt{j_q + 1/2}}\right)^2 \qquad \text{(18)}$$

where the index (m) of the product of λ-s means that the λ_m eigenvalue is omitted. The derivative $\frac{\partial D}{\partial l}$ can also be found from Equation D10 if instead of $j_\mu^{(0)} + j_\mu^{(1)}$ we write l. In the end we obtain

$$\frac{\partial D}{\partial l} = \epsilon^{p-1}(\lambda_1\lambda_2\ldots\lambda_p)^{(m)}\,J_{op+1}J_{op+2}\cdots J_{ON} \qquad \text{(D19)}$$

which together with Equations D18 and D17 gives

$$Q_n^{(2)} = -\frac{1}{2k_nJ_{on}^2}\left(\sum_{q=1}^{p}\int_{r_0}^{\infty}\psi_{on}W^1_{nq}\psi_{oq}dr\,\frac{T_{qn}}{A_q\sqrt{j_q + 1/2}}\right)^2 \qquad \text{(D20)}$$

REFERENCES

1. **Tabor, M.,** The onset of chaotic motion in dynamical systems, *Adv. Chem. Phys.*, 46, 73, 1981.
2. **Casati, G., Chirikov, B. V., and Ford, J.,** Marginal local instability of quasi-periodic motion, *Phys. Lett.*, 77A, 91, 1980.
3. **Slater, N. B.,** *Theory of Unimolecular Reactions*, Cornell University Press, Ithaca, N.Y., 1959
4. **Rutherford, E.,** Scattering of α and β particles by matter and structure of atom, *Philos. Mag.*, 21, 669, 1911.
5. **Ford, K. W. and Wheeler, J. A.,** Semiclassical description of scattering, *Ann. Phys. (N.Y.)*, 7, 259, 1959.
6. **Berry, M. V. and Mount, K. E.,** Semiclassical approximations in wave mechanics, *Rep. Prog. Phys.*, 35, 315, 1972.
7. **Feynman, R. P. and Hibbs, A. R.,** *Quantum Mechanics and Path Integrals*, McGraw-Hill, New York, 1965.
8. **Miller, W. H. and George, T. F.,** Semiclassical theory of electronic transitions, *J. Chem. Phys.*, 56, 5637, 1972.
9. **Wigner, E. P.,** Lower limit of energy derivative of scattering phase shift, *Phys. Rev.*, 98, 145, 1955.
10. **Smith, F. T.,** Lifetime matrix in collision theory, *Phys. Rev.*, 118, 349, 1960.
11. **Goldstein, H.,** *Classical Mechanics*, Addison-Wesley, Reading, Mass., 1959, Chap. 10.5.
12. **Baz', A. I., Zel'dovich, Y. B., and Peremelov, A. M.,** Scattering reactions and decay in nonrelativistic quantum mechanics, Israel Program for Scientific Translations, Jerusalem, 1969, chap. 6.
13. **Newton, R. G.,** *Scattering Theory of Waves and Particles*, McGraw-Hill, New York, 1966, chap. 19.
14. **Watson, G. H.,** *Proc. R. Soc.*, A95, 83, 1918.
15. **Regge, T.,** Introduction to complex orbital momenta, *Nuovo Cimento*, 14, 951, 1959.
16. **Alfaro, V. and Regge, T.,** *Potential Scattering*, North-Holland, Amsterdam, 1965.
17. **Collins, P. D. B.,** *Regge Theory and High Energy Physics*, Cambridge University Press, Cambridge, 1977.
18. **Vezzetti, D. J. and Rubinow, S. I.,** Asymptotic solution of the Schrödinger equation, *Ann. Phys. (N.Y.)*, 35, 373, 1965.
19. **Shuler, K. E. and Zwanzig, R.,** Quantum mechanical calculation of harmonic oscillator transition probabilities, *J. Chem. Phys.*, 33, 1778, 1960.
20. **Kleinman, B. and Tang, K. T.,** Comparison of quantum and classical theories of an idealized three-body rearrangement collision, *J. Chem. Phys.*, 51, 4587, 1969.
21. **Born, M. and Wolf, E.,** *Principle of Optics*, Pergamon Press, Oxford, 1975, chap. 3.2.
22. **Maslov, V. P. and Fedoriuk, M. V.,** *Semiclassical Approximation in Quantum Mechanics*, D. Reidel, Boston, 1981.
23. **Fonda, L. and Newton, R. G.,** K-hyperon production threshold phenomena, *Nuovo Cimento*, 14, 1027, 1959.
24. **Van Beveren, E., Dullemond, C., Rijken, T. A., and Rupp, G.,** Analytically solvable multichannel Schrödinger model for hadron spectroscopy, in *Resonances-Models and Phenomena, Lecture Notes in Physics 211*, Albeveria, S., Ferreira, L. S., and Streit, L., Eds., Springer-Verlag, Berlin, 1984, 182.
25. **Dullemond, C. and Van Beveren, E.,** Confining potentials and Regge poles, *Ann. Phys. (N.Y.)*, 105, 318, 1977.
26. **Morse, P. H. and Feshbach, H.,** *Methods of Theoretical Physics*, McGraw-Hill, New York, 1953, chap. 4.8.
27. **Nussenzveig, H. M.,** High-frequency scattering by impenetrable sphere, *Ann. Phys. (N.Y.)*, 34, 23, 1965.
28. **Bosanac, S.,** Calculation of scattering amplitude in Regge representation, *Mol. Phys.*, 35, 1057, 1978.
29. **Rubinov, S. I.,** Scattering from a penetrable sphere at short wave-lengths, *Ann. Phys. (N.Y.)*, 14, 305, 1961.
30. **Sommerfeld, A.,** *Partial Differential Equations of Physics*, Academic Press, New York, 1949.
31. **Nussenzveig, H. M.,** High-frequency scattering by transparent sphere, I, *J. Math. Phys.*, 10, 82, 1969.
32. **Nussenzveig, H. M.,** High-frequency scattering by transparent sphere, II, *J. Math. Phys.*, 10, 125, 1969.
33. **Newton, R. G.,** *The Complex J-Plane*, Benjamin, New York, 1964.
34. **Remler, E. A.,** Complex angular momentum analysis of atom-atom scattering experiments, *Phys. Rev.*, A3, 1949, 1971.
35. **Sukumar, C. V., Lin, S. L., and Bardsley, J. N.,** Regge poles and Watson-Sommerfeld transformation in low energy atomic collisions, *J. Phys.*, B8, 577, 1975.
36. **Connor, J. N. L.,** Regge pole analysis of large angle scattering, *Mol. Phys.*, 29, 745, 1975.
37. **Jost, R.,** Über die falschen Nullstellen der Eigenwerte der S-matrix, *Helv. Phys. Acta*, 20, 256, 1947.
38. **Newton, R. G.,** Structure of many-channel S-matrix, *J. Math. Phys.*, 2, 188, 1961.
39. **Newton, R. G.,** *Scattering Theory of Waves and Particles*, McGraw-Hill, New York, 1966.
40. **Fonda, L.,** Bound states embedded in the continuum, *Ann. Phys. (N.Y.)*, 22, 123, 1963.

41. **Bosanac, S.,** Perturbation expansion of multichannel Regge poles, *J. Math. Phys.*, 21, 1881, 1980.
42. **Newton, R. G.,** Inelastic scattering, *Ann. Phys. (N.Y.)*, 4, 29, 1958.
43. **Bosanac, S.,** Semiclassical perturbation expansion of multichannel scattering matrix, *J. Math. Phys.*, 20, 396, 1979.
44. **Vedder, H. J., Chawla, G. K., and Field, R. W.,** Observation of quasibound energy levels and tunnelling lifetimes in $Na_2B^1\pi_0$ state, *Chem. Phys. Lett.*, 111, 303, 1984.
45. **Connor, J. N. L.,** On semi-classical description of molecular orbiting collisions, *Mol. Phys.*, 15, 621, 1968.
46. **Delos, J. B. and Carlson, C. E.,** Semiclassical calculation of Regge poles, *Phys. Rev.*, A11, 210, 1975.
47. **Connor, J. N. L., Delos, J. B., and Carlson, C. E.,** On Regge trajectories for ion-atom and atom-atom potentials, *Mol. Phys*, 31, 1181, 1976.
48. **Bosanac, S.,** Zero-angular momentum resonances in atom-atom collisions, *Phys. Rev.*, A28, 1344, 1983.
49. **Bosanac, S.,** Time-delay analysis of zero-angular momentum resonances, *Phys. Rev*, A30, 153, 1984.
50. **Buck, U.,** Elastic scattering, *Adv. Chem. Phys.*, 30, 313, 1975.
51. **Breit, G. and Wigner, E. P.,** Capture of slow neutrons, *Phys. Rev.*, 49, 519, 1936.
52. **Breit, G.,** Theory of resonance reactions, in *Handbuch der Physik*, Vol. XLI/1, Flügge, S., Ed, Springer-Verlag, Berlin, 1959, chap. 4.
53 **Pauly, H.,** Elastic collisions, in *Atom-Molecule Collision Theory*, Bernstein, R. B., Ed., Plenum Press, New York, 1979, chap. 4.
54. **Berry, M. V.,** Uniform approximation for potential scattering involving a rainbow, *Proc. Phys. Soc.*, 89, 479, 1966
55. **Abramowitz, M. and Stegun, I. A.,** Handbook of mathematical functions, Dover, New York, 1968.
56. **Bosanac, S. and Knešaurek, K.,** Analysis of proton-He resonances in very-low-energy collisions, *Phys. Rev.*, A28, 2173, 1983.
57. **Gradsteyn, I. S. and Ryzhik, I. M.,** *Table of Integrals, Series and Products*, Academic Press, London, 1965.
58. **Fonda, L., Ghirardi, G. C., and Rimini, A.,** Decay theory of unstable quantum systems, *Rep. Prog. Phys.*, 41, 587, 1978.
59. **Gamow, G.,** Zur Quantentheorie des Atomkernes, *Z. Phys.*, 51, 204, 1928.
60. **Siegert, A. J. F.,** On derivation of description formula for nuclear reactions, *Phys. Rev.*, 56, 750, 1939.
61 **Goldberger, M. L. and Watson, K. M.,** Collision theory, John Wiley & Sons, New York, 1964, chap. 8.
62. **Wu, T.-Y. and Ohmura, T.,** *Quantum Theory of Scattering*, Prentice-Hall, London, 1962, chap. A7.
63. **Wilson, D. J. and Locker, D. J.,** Quantum vibrational transition probabilities in atom-diatomic molecule collisions, *J. Chem Phys.*, 57, 5393, 1972.
64. **Heding, J.,** *An Introduction to Phase-Integral Methods*, Methuen, London, 1962
65. **Froman, N. and Froman, P. O.,** *JWKB Approximation*, North-Holland, Amsterdam, 1965.
66. **Desai, B. R. and Newton, R. G.,** Angular momentum poles of the nonrelativistic S-matrix for spin $^1/_2$ particles, *Phys Rev.*, 129, 1437, 1963.
67. **Charap, J. M. and Squires, E. J.,** On complex angular momentum in many-channel potential scattering problems, III, *Ann. Phys (N.Y.)*, 25, 143, 1963
68 **Bosanac, S.,** Two-dimensional model of rotationally inelastic collisions, *Phys. Rev.*, A22, 2617, 1980.
69. **Bosanac, S.,** Quantum theory of atom-ellipsoid scattering in two dimensions, *Phys. Rev.*, A26, 282, 1982.
70. **Bosanac, S.,** Origin of left-right asymmetry in rotationally inelastic differential cross sections, *Phys. Rev.*, A26, 816, 1982
71. **Takayanagi, K.,** Low energy ion molecule collisions, in *Physics of Electronic and Atomic Collisions*, Datz, S., Ed, North-Holland, Amsterdam, 1982, 343.
72. **Charap, J. M. and Squires, E. J.,** On complex angular momentum in many channel potential scattering problems, I, *Ann. Phys.*, 20, 145, 1962.
73. **Charap, J. M. and Squires, E. J.,** On complex angular momentum in many channel potential scattering problems, II, *Ann. Phys*, 21, 8, 1963.
74 **Calogero, F., Charaz, J. M., and Squires, E. J.,** Continuation in total angular momentum of partial wave scattering amplitudes for particles with spin, *Ann. Phys.*, 25, 325, 1963.

INDEX

F

G

H

I

J

K

L

M

N

O

P

Q

R

S